NOx, PM and CO_2 Emission Reduction in Fuel Combustion Processes

NOx, PM and CO_2 Emission Reduction in Fuel Combustion Processes

Editors

Monika Kosowska-Golachowska
Tomasz Czakiert

MDPI

Basel • Beijing • Wuhan • Barcelona • Belgrade • Novi Sad • Cluj • Manchester

Editors

Monika Kosowska-Golachowska
Department of Thermal Machinery, Faculty of Mechanical Engineering and Computer Science
Czestochowa University of Technology
Czestochowa, Poland

Tomasz Czakiert
Department of Advanced Energy Technologies, Faculty of Infrastructure and Environment
Czestochowa University of Technology
Czestochowa, Poland

Editorial Office
MDPI
St. Alban-Anlage 66
4052 Basel, Switzerland

This is a reprint of articles from the Special Issue published online in the open access journal *Energies* (ISSN 1996-1073) (available at: https://www.mdpi.com/journal/energies/special_issues/nox_pm_co2_emission_reduction_combustion_processes).

For citation purposes, cite each article independently as indicated on the article page online and as indicated below:

Lastname, A.A.; Lastname, B.B. Article Title. *Journal Name* **Year**, *Volume Number*, Page Range.

ISBN 978-3-0365-8990-9 (Hbk)
ISBN 978-3-0365-8991-6 (PDF)
doi.org/10.3390/books978-3-0365-8991-6

Contents

energies

MDPI

Article

Physicochemical Analysis of the Particulate Matter Emitted from Road Vehicle Engines

Remigiusz Jasiński [1], Beata Strzemiecka [2], Iwona Koltsov [3], Jan Mizeracki [3] and Paula Kurzawska [1,*]

[1] Faculty of Civil and Transport Engineering, Poznan University of Technology, 60-965 Poznan, Poland; remigiusz.jasinski@put.poznan.pl
[2] Institute of Chemical Technology and Engineering, Poznan University of Technology, 60-965 Poznan, Poland; beata.strzemiecka@put.poznan.pl
[3] Laboratory of Nanostructures, Institute of High-Pressure Physics of the Polish Academy of Sciences, 01-142 Warsaw, Poland; i.koltsov@labnano.pl (I.K.); j.mizeracki@labnano.pl (J.M.)
* Correspondence: paula.kurzawska@put.poznan.pl

Abstract: Air pollution with particulate matter from transport sources is a serious problem in terms of air quality and its impact on human health. The article attempts to test the emitted particles from piston engines in terms of their physical properties and chemical composition. The research test objects were a diesel engine with Euro 5 emission class and a petrol engine, which was a part of the scooter drive system. The conducted research consisted in the analysis of the number, mass, and volume of particles, as well as chemical analysis, using the methods: Fourier Transform Infrared Spectroscopy (FTIR), Scanning Electron Microscope with Energy Dispersive Spectroscopy (SEM-EDS), and Evolved gas analysis (EGA). The diesel engine emitted particles in the range of 50–120 nm. With the increase in the engine load, the specific emission of particulate matter increased. In the case of a gasoline engine running without load, the emission of particles smaller than 30 nm was mainly observed. Increasing load of the gasoline engine resulted in an increase in both the concentration of particles and their diameter (average diameter to 90 nm). FTIR analysis showed higher black carbon content in the case of the sample taken from the diesel engine. In order to carry out a more detailed chemical analysis, the EGA and SEM methods were used. On their basis, the chemical composition of particles was presented, and a greater ability to agglomerate of a gasoline engine particles was found.

Keywords: combustion engines; particles emissions; chemical composition; SEM-EDS; FTIR

Citation: Jasiński, R.; Strzemiecka, B.; Koltsov, I.; Mizeracki, J.; Kurzawska, P. Physicochemical Analysis of the Particulate Matter Emitted from Road Vehicle Engines. *Energies* **2021**, *14*, 8556. https://doi.org/10.3390/en14248556

Academic Editor: Islam Md Rizwanul Fattah

Received: 9 November 2021
Accepted: 14 December 2021
Published: 18 December 2021

1. Introduction

Air quality has improved significantly in European cities over the past 20 years. The emission of sulfur oxides was mainly reduced: ~80%, nitrogen oxides: ~46%, non-methane volatile organic compounds: ~44%, particles with a diameter lower than 2.5 μm and 10 μm: 30% [1]. Nevertheless, European society is exposed to air pollution by $PM_{2.5}$ and O_3 at levels significantly exceeding the WHO limit values for the protection of human health [1]. This is related to premature mortality [2], estimated at 4.5 million deaths worldwide and approximately 0.5 million in Europe annually [3]. WHO, on their official website, states that 91% of people live in the area where air quality levels exceed WHO limits [4]. WHO, in Reference [5], says that the level of PM2.5 does not have a significant influence on human health in long-term time if it is 10 μg/m^3 per year. Today, according to the official data presented by WHO, the level of PM2.5 in all European countries is 25 μg/m^3 per year (except Switzerland, where it is 10 μg/m^3 per year) [6].

Exposure to particles with diameter less than 2.5 μm for several hours may result in cardiovascular-related mortality. Long-term exposure (e.g., several years) increases the risk of cardiovascular mortality, can cause pulmonary inflammation, and reduces life expectancy in the more vulnerable segments of the population by several months to several years [7,8]. Atmospheric pollution with particles emitted by combustion engines adversely affects human health. Particles smaller than 1 μm in diameter can cause various diseases, especially heart and lung diseases, and related deaths. The worsening of disease is associated with the long-term effects of particles in the environment. They contribute to diseases, such as asthma and bronchitis. They are also one of the causes of heart arrhythmias and heart attacks. The most serious problems arise from the interaction of fine particles. People with heart and lung diseases, the elderly, and children have the lowest resistance to the negative effects of particulate matter. In impact analysis of air pollution on humans, in addition to common diseases, one should consider human cognitive abilities. Some scientists argue that air pollution negatively affects cognitive performance [9–11]. It was found that there are significant negative relationships between air pollutant exposure and cognitive function, especially PM2.5 exposure and long-term exposure. Additionally, it was found that the effect of PM on cognitive traits differs depending on the characteristics and attributes of the exposed person [12].

One of the main sources of particulate matter emissions in cities is road transport, especially vehicles with internal combustion engines (ICE), which are one of the main sources of particles emissions smaller than 1 μm [13,14]. Particles emissions emitted by compression ignition engines are from 6 to 10 times higher than emitted from spark ignition engines [15]. According to European Automobile Manufacturers Association (ACEA) Report for 2019, in European Union, 41.9% of passenger cars, 91.2% of light commercial vehicles and 98.3% of medium and heavy commercial vehicles were diesel vehicles [16]. The average retirement age of diesel cars in 2012–2013 was 14.0 years, and it increased by almost 12% since 2006. Lifetime mileage of diesel cars was approximately 208,000 km, based on data from 2012–2013 [17].

Particulate matter is one of the four main pollutants emitted by diesel engines, along with carbon monoxide (CO), nitrogen oxides (NOx), and hydrocarbons (HC). Particulate matter in internal combustion engines mainly arises as a result of incomplete combustion of hydrocarbons contained in the fuel and also from lube oil [15]. More than 90% of particles emitted by diesel engine are smaller than 1 μm, and most of the particles are between 15 and 40 nm in diameter [15,18]. Size distribution of particles emitted from diesel engines is bimodal, with mainly nucleation and accumulation mode [19]. There are many factors that influence the amount of particulate emissions from diesel engines. These are, among others: the age of the engine, the properties of the fuel supplying the engine, the level of maintenance of the vehicle and the driving style, environmental conditions, and the exhaust gas treatment technology used, but also the amount of oil scraped into the combustion chamber through the piston rings [14,20]. In order to reduce particulate emissions from internal combustion engines, particulate filters, such as the diesel particulate filter (DPF), are used. To function properly, filters used in diesel engines require periodic regeneration, which allows to remove accumulated solid particles. The efficiency of particle reduction for a properly functioning diesel particulate filter is about 95% [21]. In 2009, Europe established a new EURO 5 standard, and, since then, DPF is mandatory in new diesel cars [22]. In 2011, a new EURO 5b standard came into force for light passenger and commercial vehicles, which set a limit for the number of particles of 6 × 1011 per km, and was the first limitation of particles number [23]. The last established standard is EURO 6d, which has been in force since January 2021, but the limits for particles number and mass have not changed since EURO 5. As a diesel particulate filter is mandatory on new cars, many users of diesel cars remove the DPF to avoid the cleaning costs [22]. As a result of the removal of the DPF, the car no longer meets the applicable standards and is very harmful to the environment in terms of particulate matter. In many European countries, removal of DPFs is forbidden and punished with high fines [22,24].

One of the solutions to reduce pollutant emissions in cities is to completely ban or limit the entry of vehicles that do not meet the relevant standards. Currently, there are Low Emission Zones (LEZs) in many countries, such as Italy, Spain, France, Germany, Belgium, and others, where the most polluting vehicles are regulated [25]. More and more countries are considering a permanent ban on diesel and gasoline cars to favor electric cars. Thirteen countries and several regions and cities announced ICE car bans until 2045 [26,27]. For this reason, development and researches of new technologies and trends in powertrains is inevitable. Currently, there are few configurations of electric powertrains: Plug-in Electric Vehicle (PEV), Plug-in Hybrid Vehicle (PHEV), Battery electric vehicle (BEV), and Extended-Range Electric Vehicle (EREV) [28–30]. Thanks to an electric or hybrid drive system, that does not generate exhaust gases during use, it is possible to reduce greenhouse gases and other harmful exhaust gas compounds arising during combustion in a conventional engine. While electric cars still need electricity, which, in many countries, is produced from fossil fuels, reduction of harmful exhaust gas compounds emission is not so significant as it can be when electricity is produced from renewable energy sources [31]. However, another alternative for combustion engines is the Hydrogen Powered Vehicle (FCV), which use hydrogen fuel, that can be produced from zero-emission energy source [31]. Alternative powertrains can be considered as a relatively quick solution for reduction of harmful emission from road transport in countries with well-developed infrastructure for electric cars, while, in many countries, the main barriers will be the lack of charging infrastructure, insufficient or lack of information and promoting programs on the national level, and also high price for electric powertrains [30].

The other problem of PM is the influence of the chemical composition on the human health. There are many questions and problems that concern the analysis of the chemical composition of the PM:

1. Very complex composition of PM: many organic and inorganic compounds were found in PM up-to-date [32,33].
2. Very small amount of the compounds that are at the limit of the detection of the known nowadays techniques (e.g., benzo(*a*)pyrene present at PM2.5 was found equal to 15.4 ng/m^3 per year in the Polish city of Katowice).
3. Many factors influence human health, and it is extremely hard (even impossible) to prove that the higher level of some particular compound is dangerous.
4. Because of the small amount of PM, the measurements are performed in the long term. Most often, the chemical composition of PM is assessed in the measurements at least collected in a day, mostly one month, or year. Thus, most often, the concrete hazardous compound present in PM can be assessed in the long term, and what is important for the risk for human is the time of exposure that is extremely (not possible) to assess as it will be very variable and dependent on many factors (e.g., weather, humidity).

Generally, WHO states that the largest sources of PM emission are solid fuel, households, and transport.

In this paper, the authors take into account the transport factor and assess the influence of the engine type, the conditions of its work on PM amount emission, and chemistry. Compared to other studies that have been carried out in the past few years, this article includes both particulate matter data analysis obtained from the particles analyzer, but also three different chemical analyzes: Fourier Transform Infrared Spectroscopy (FTIR), Scanning Electron Microscope with Energy Dispersive Spectroscopy (SEM-EDS), and Evolution Gas Analysis (EGA), while most research focuses on only one chemical analysis [34–36]. The authors of this paper verify the hypothesis of whether the type of engine and its work conditions influence/does not influence the chemistry of PM and this chemistry difference are/are not significant. Moreover, this paper gives a brief answer as to what technique can be adequate for quick and accurate PM chemistry studying.

2. Materials and Methods

2.1. Test Engine, Fuels, Operating Schedule

One of the tested engine was a compression ignition engine made by FCE Powertrains. The main parameters of this engine are presented in Table 1. The arrangement of the cylinders is in line with the Double Overhead Camshaft. The engine is fueled with diesel fuel and meets the EURO 5 standard. The engine has Common Rail direct injection system. The engine also has anelectronically controlled Exhaust Gas Recirculation valve, the operation of which is aimed at reducing the emission of harmful compounds in the exhaust gas. The EGR valve directs some of the exhaust gas from the exhaust system to the engine's intake system, thereby lowering the temperature in the combustion chamber and slowing down the combustion process. This allows the emission of nitrogen oxides to be reduced. The engine is also equipped with an oxidation catalytic converter and Diesel Particulate Filter.

Table 1. Engine's parameters.

Parameter	Value
Engine type	Diesel engine
Capacity	1.3 dm^3
Power	55 kW for 4000 rpm
Torque	190 Nm for 1750–2500 rpm
Number of cylinders	4
Number of valves	16
Piston stroke	82 mm
Cylinder diameter	69.6 mm
Compression ratio	17.0:1

The second tested engine was a spark ignition four-stroke engine with a capacity of 0.05 dm^3 used in motor scooter. The engine is fueled with unleaded gasoline 95. Diesel engines were tested on an engine dynamometer test cell equipped with a brake AMX-210/100. The maximum power of the brake is 100 kW, maximum torque is 240 Nm, and the rotation speed is 10,000 rpm. The tests were based on the engine load characteristics, which consist of assessing the change in basic engine parameters, depending on its load at constant crankshaft rotation speed. The tests were made for three rotation speeds: 1800, 2100, and 2400 rpm. For every rotation speed, at least seven engine loads were tested, from 15 Nm to 135 Nm, in steps of 20 Nm. The exception is the test for 2100 rpm, when the regeneration of DPF has become apparent, and, for this rotation speed, the number of tested loads is 8 points.

In the case of the petrol engine, the research was conducted differently. Due to the fact that the engine was mounted on a scooter, it was necessary to use a chassis dynamometer for two-wheelers. Due to the above fact, it was only possible to carry out tests for the qualitative determination of whether the engine was under load or not.

2.2. *Apparatus and Procedures*

2.2.1. PM Emissions

Particulate matter concentration was measured with an Engine Exhaust Particle Sizer (EEPS 3090 analyzer, Shoreview, MN, USA). EEPS enables performance of the particle size distribution for discrete range of particle diameters, from 5.6 nm to 560 nm. The exhaust gases are directed to the device via dilution system and temperature maintenance system. The pre-filter collects particles larger than 1 μm in diameter that are outside the measuring range devices [37,38].

During the tests, the number concentration of particles in the exhaust gas was measured by EEPS, and the particulate matter sample was collected separately on the filter. The filter used in test was a Glass Fiber Filter 37 mm. For the diesel engine, the EEPS probe was behind the diesel particulate filter, while the Glass Fiber Filter was in front of diesel particulate filter. The filter was placed in a separate pipe in which the pump drew in the exhaust gases from the exhaust system. For the petrol engine, the EEPS probe and the filter were in the exhaust of the motor scooter, and the filter also was between the exhaust system and the pump, which drew the exhaust gases in to pass them through the filter.

2.2.2. Fourier Transform Infrared Spectroscopy (FTIR)

FTIR spectra of PM adsorbed on the filter were collected using Vertex 70, produced by Bruker, Billerica, MA, USA. The PM adsorbed on filter without any treatment were analyzed by use of the Attenuated total reflection (ATR) technique. The PIKE MIRacle™ accessory with diamond crystal was used. The measurements were performed at a resolution of 0.5 cm^{-1} in the wavenumber range 4000–400 cm^{-1}.

2.2.3. Scanning Electron Microscope with Energy Dispersive Spectroscopy (SEM-EDS)

The morphology of the filters was investigated using a scanning electron microscope ULTRA PLUS (ZEISS, Oberkochen, Germany). The quantitative X-ray microanalysis of the materials was performed by the energy-dispersive spectrometry method using an X-ray spectrometer (Quantax 400, Bruker, Billerica, MA, USA).

2.2.4. Evolved Gas Analysis (EGA)

Evolved gas analysis combining differential scanning calorimetry (DSC) and thermogravimetry analysis (TG) (STA 449 F1 Jupiter apparatus from Netzsch GmbH, Selb, Germany) was conducted to determine the thermal decomposition of filters. Simultaneously, to the thermal decomposition of filters, the analysis of released gases took place using a mass spectrometer (QMS 403 Aeolos, from Netzsch GmbH, Selb, Germany). The measurements were carried out in synthetic air (10 cm^3/min) at heating rate of 10 °C/min over a temperature range of 25–1200 °C. The QMS was operated with an electron impact ionizer with an energy of 70 eV. The mass/charge (m/z) ratio was recorded in the range 2–200 amu during the measurements. For the experiment purpose, the biopsy tool was used to cut circular shape from the filter. In the test, 3 of those shapes were measured stacked on each other. The initial total sample weight taken for experiment was approximately 6 mg.

3. Results

3.1. Particle Number Emission Indices

The average PM number emission indices (EI_N) and their associated error bars were calculated from EEPS data, as provided in Figure 1, for the different rotation speed. The average EI_N for 1800 rpm decreased from 1.9×10^{15} to 9.3×10^{13} particles per kilogram of fuel as the engine torque increased from 15 to 55 Nm. From a 55 Nm average, EI_N increased to 5.9×10^{15} for 95 Nm, which was the highest EI_N for 1800 rpm. The figure shows that, in comparison to the other rotation speeds, 1800 rpm had the highest EI_N values over the entire range of engine torque. For 2100 rpm rotation speed, EI_N values decreased from 8.9×10^{13} to 6.8×10^{13} particles per kilogram of fuel as the engine torque increased from 15 to 115 Nm. About torque equal to 144 Nm regeneration of DPF (diesel particulate filter) has become. It caused an increase of EI_N value from 6.9×10^{13} to 4.9×10^{14} particles per kilogram as the engine torque increased from 115 to 144 Nm. The average EI_N for 2400 rpm is after regeneration of DPF; thus, the average values of particles per kilogram of fuel are significantly lower than for 1800 and 2100 rpm. The values of EI_N decreased from 1.9×10^{13} to 1.0×10^{13} as the engine torque increased from 15 to 135 Nm.

3.2. PM Mass Emission Indices

The PM mass emission indices (EI_M), under various test conditions, were calculated from the EEPS measurements. Figure 2 shows the effect of rotation speed and engine torque on the EI_M. Similar to particle number emission indices, the mass emission indices in comparison to other rotation speeds are the highest for 1800 rpm rotation speed in the entire range of engine torque. For 1800 rpm, the average EI_M decreased from 1.14×10^3 to 3.0×10 milligrams per kilogram of fuel as the engine torque increased from 15 to 55 Nm. After that point, it increased to the highest value of 1.23×10^3 for 95 Nm. For 2100 rpm rotation speed, the EI_M decreased from 6.06×10 to 7.73 milligrams per kilogram of fuel as the engine torque increased from 15 to 75 Nm. From this point, it increased to 2.94×10^2 for 144 Nm, while the regeneration of DPF started. PM mass emission indices for rotation speed

2400 rpm were the lowest from all tested rotation speeds because of the DPF regeneration. The average value of EI_M decreased from 6.75 to 2.91 milligrams per kilogram of fuel as the engine torque increased from 15 to 135 Nm.

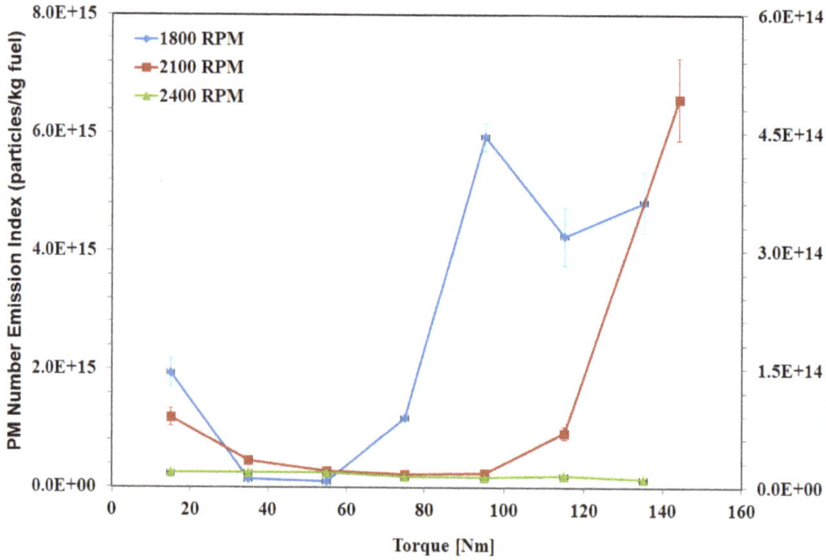

Figure 1. Particle number emissions index (EI_N) versus torque. Error bars represent single standard deviation.

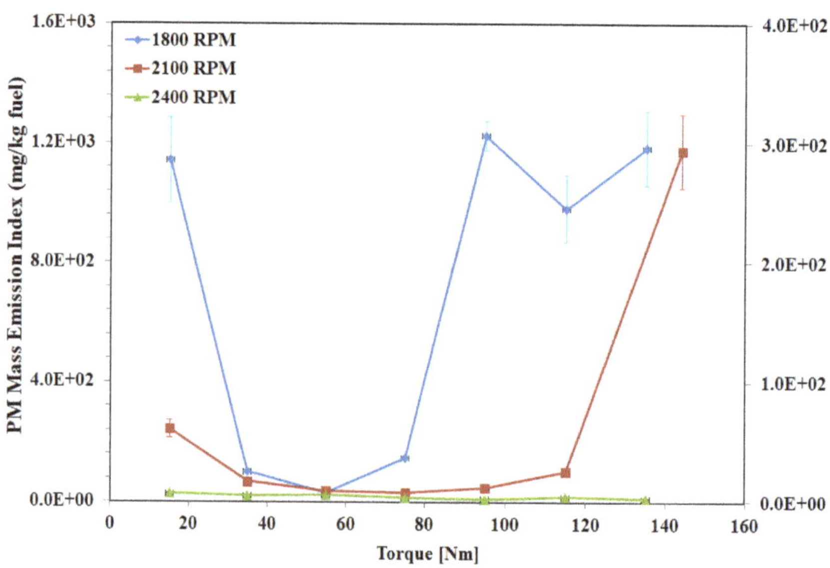

Figure 2. PM mass emission index (EI_M) versus torque. Error bars represent one standard deviation.

3.3. Particle Size Distribution

The data recorded by the EEPS was averaged for the same engine torque and then converted to differential number-based ($dEI_N/dlogD_p$) and differential volume-based ($dEI_V/dlogD_p$) particle size distributions. Figure 3 represents example plots of $dEI_N/dlogD_p$

and $dEI_V/dlogD_p$ under different engine torques for rotation speed 1800 rpm (Figure 3a,b), 2100 rpm (Figure 3c,d), and for 2400 rpm (Figure 3e,f).

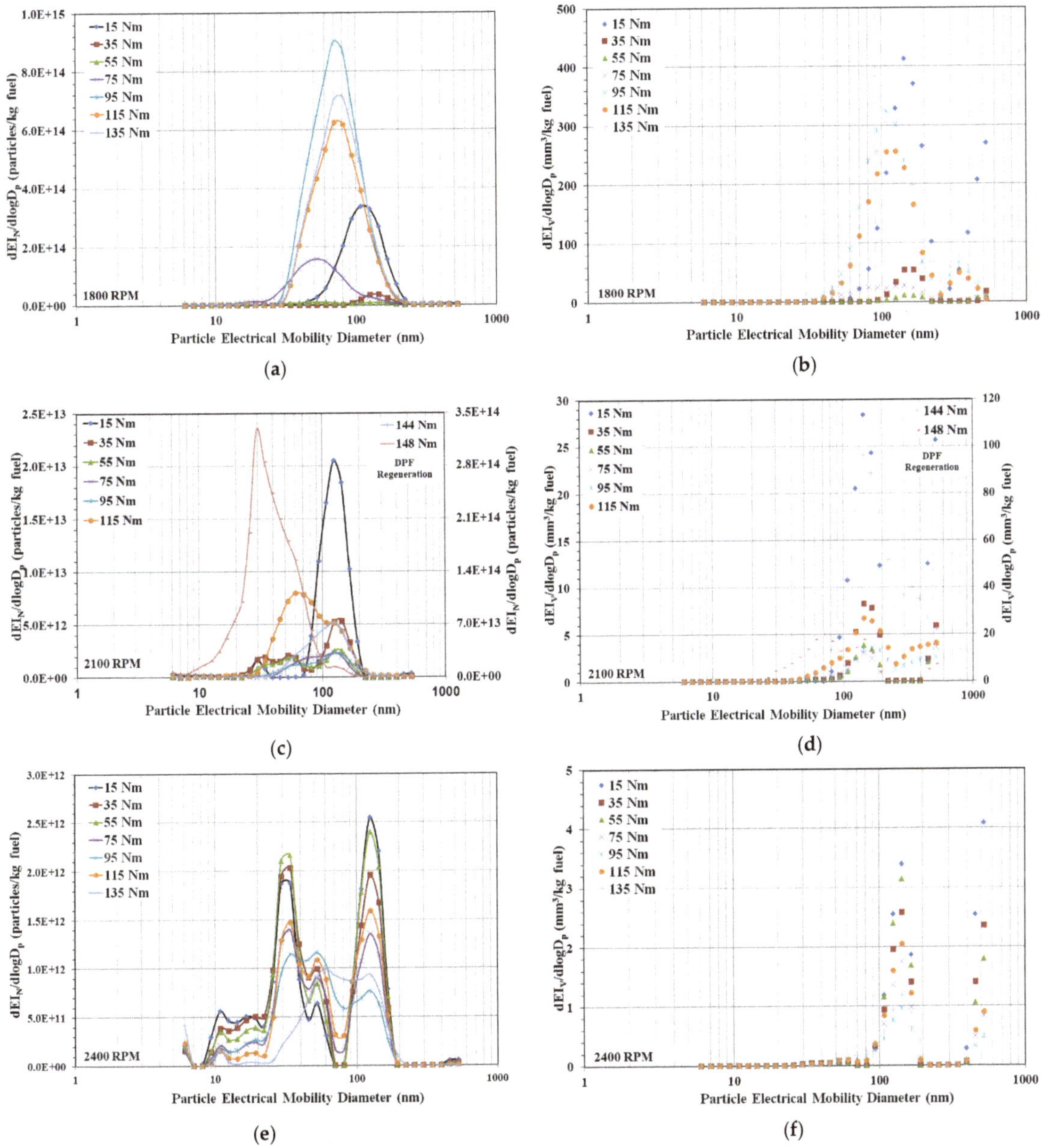

Figure 3. Differential particle number emission index (EI_N) and differential particle volume emission index (EI_V) PSDs: (**a**) EI_N PSD for 1800 rpm; (**b**) EI_V PSD for 1800 rpm; (**c**) EI_N PSD for 2100 rpm; (**d**) EI_V PSD for 2100 rpm; (**e**) EI_N PSD for 2400 rpm; (**f**) EI_V PSD for 2400 rpm.

For rotation speed 1800 rpm, the particles formed in the accumulation mode dominate; thus, number-based PSD exhibited a single mode log-normal distribution. EI_N and EI_V

was the highest for the last engine torques (95, 115, and 135 Nm), but also for the first tested engine torque (15 Nm).

For 2100 rpm, mostly, the particles formed in the accumulation mode dominate. When the torque reached the point of 144 Nm, regeneration of DPF started; thus, for 148 Nm, the particles formed in nucleation mode dominate. Particle number emission index before regeneration of DPF is the highest for first tested torque (15 Nm) and for the last tested engine torque before DPF regeneration (115 Nm). After DPF regeneration started, the EI_N is the highest for 148 Nm. Particle volume emission index before DPF regeneration is the highest for 15 Nm, and, for 144 Nm, when the regeneration happened.

PSD for 2400 rpm shows number emission index and volume emission index after regeneration of DPF. Number-based PSD exhibited a multimodal log-normal distribution. The particles formed in accumulation mode dominated, but there were also particles formed in nucleation mode. In addition, peak of accumulation mode was higher than nucleation mode. The highest EI_N and EI_V for almost all particles diameters was for the three first tested engine torques (15, 35, and 55 Nm). The center of nucleation and accumulation peaks mostly decreased as engine torque increased. The median of change in average particle number and mass emission compared to 1800 rpm rotational speed was, respectively, 99.0% and 99.4%.

Figures 4–6 show the measurements results of particulate matter from the scooter in two phases: idling and under an average load on a chassis dynamometer. Figure 4 shows the dimensional distribution of particles. During idling, particles with a diameter of about 10 nm dominate, reaching the maximum value of 4.5×10^5. The engine load caused the dimensional distribution to shift to the right, towards larger particles. Thus, the dominant diameter is approximately 80 nm. In addition to increasing the mean particle size, an increase in the concentration of the particles number was noted, reaching 1.3×10^6 for the diameter dominating in the distribution.

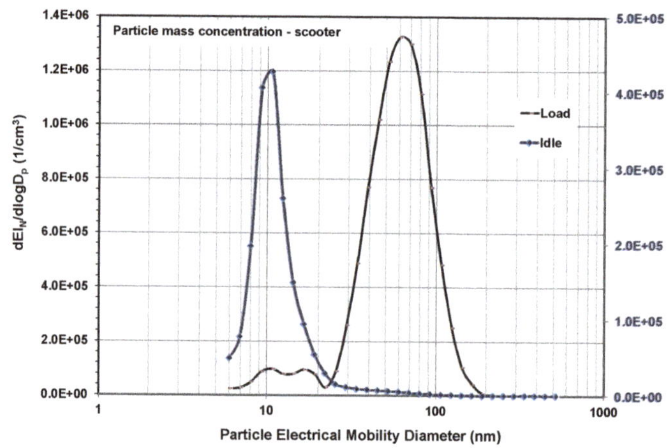

Figure 4. Particle number distribution obtained for scooter.

Figure 5 shows the mass distribution of particles. During idling, a trimodal mass distribution was noted. The largest share in the emitted mass particles is represented by particles with diameters greater than 50 nm. A significant share is attributed to particles with a diameter of approximately 10 nm, which results from their dominant number in the dimensional distribution. The engine load significantly changed the mass distribution to the unimodal one. It shows that over 95% of the emitted mass of particles were caused by particles with a diameter of 30–200 nm. Additionally, a significant increase in mass concentration was noted.

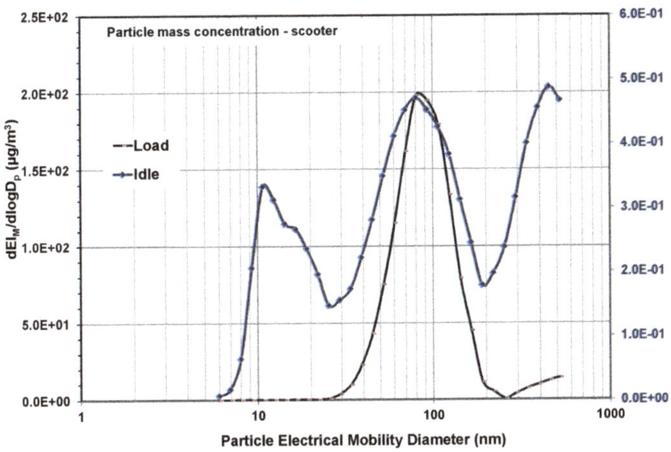

Figure 5. Particle mass distribution obtained for scooter.

Figure 6 shows the volume distribution of particles. When the engine is idling, the majority of the emitted volume of particles correspond to particles with a diameter of about 10 nm, the number of which is the largest. Particles with diameters of 30–200 nm also have a significant share in the emitted volume. In the case of engine operation under load, it was found that particles with diameters of 30–200 nm are responsible for the vast majority of the generated volume of particles.

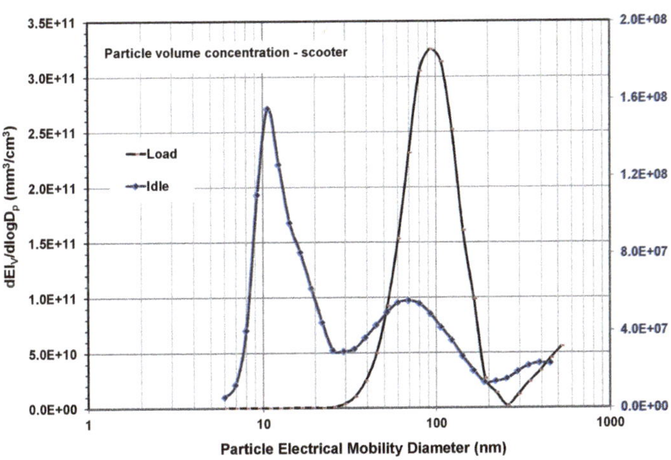

Figure 6. Particle volume distribution obtained for scooter.

3.4. FTIR ATR of PM Adsorbed on Filter

ATR spectra of PM adsorbed on the filter are presented in Figure 7.

All adsorbed PMs collected from both engines (Diesel and petrol) possess carbon black: a baseline sloping down to the right (%T display) as carbon black displays absorption over the entire region from 4000 to 400 cm^{-1}, and, when using the ATR technique, the effect of carbon black becomes greater, with deeper light penetration at the long wavelength (low wavenumber) end. The largest amount of carbon black is present at PMs emitted from Diesel engine, both working with and without load and during cold start. Slightly less carbon black is emitted in the case of an engine without load than with load. There are no evidence signals from alkanes or aromatic compounds; however, the weak signals at ~2900 cm^{-1} at the range 1650–1400 cm^{-1} might be seen. It is larger for PM from the Diesel engine than petrol engine. The signals at ~2300 cm^{-1} and ~2100 cm^{-1} can be from nitryl compounds or alkanes, and they are stronger in the case of PMs from the Diesel engine than petrol engine. During the cold start of engine, more carbon black is emitted that is stronger for the Diesel engine than petrol engine. To summarize, the PMs from different engines, and working under the different conditions, emitted PMs that differ in the chemical composition.

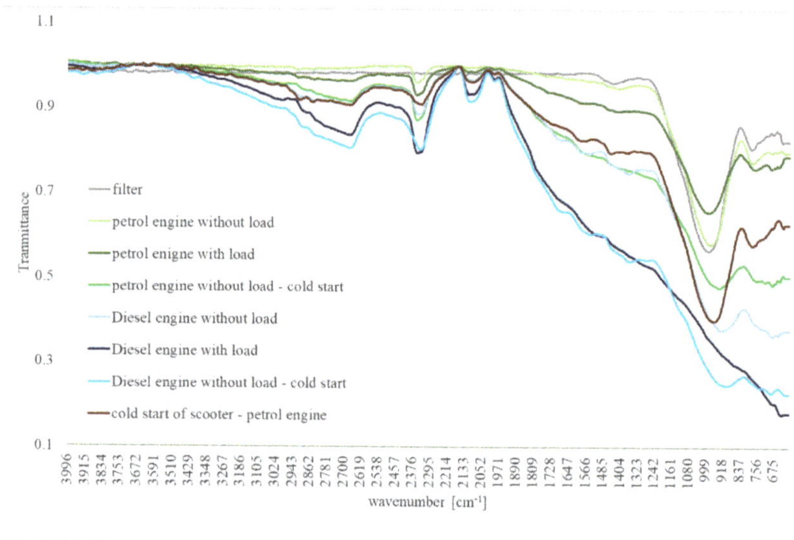

Figure 7. FTIR ATR spectra of PM adsorbed on filter from Diesel and petrol engine.

FTIR technique showed similarity and difference of the PM chemistry coming from the different engines and during their different work; however, this does not give an exact answer about the chemical composition of PM. More detailed and sophisticated methods must be applied. For this reason, in this study, the EGA technique was used. The aforementioned techniques were used in this research to investigate the thermal behavior of the filters and, at the same time, to examine the gasses released during thermal treatment. As a reference sample, the pure filter was investigated, and the results are shown in Figure 8.

3.5. *Evolved Gas Analysis (EGA)*

It can be seen in Figure 8a that decomposition of the pure filter is gradual with the total weight loss of approximately 10 wt.%. Analysis of released gasses during decomposition showed presence of CO_2 $m/z = 44$ and moisture $m/z = 18$ (Figure 8b, Table 2), where the first signal was present from approximately 200 to 600 °C, and second one from approximately 200 to 900 °C, respectively.

The results for petrol engine are shown in Figure 9. From the room temperature up to approximately 600 °C, exothermic combustion of the filter is visible (Figure 9a). This process is associated with ~8 wt.% mass loss and release of some organic compound/s (C_2H_2S or $C_2H_5CHNH_2$ ($m/z = 58$) and water (Table 2)).

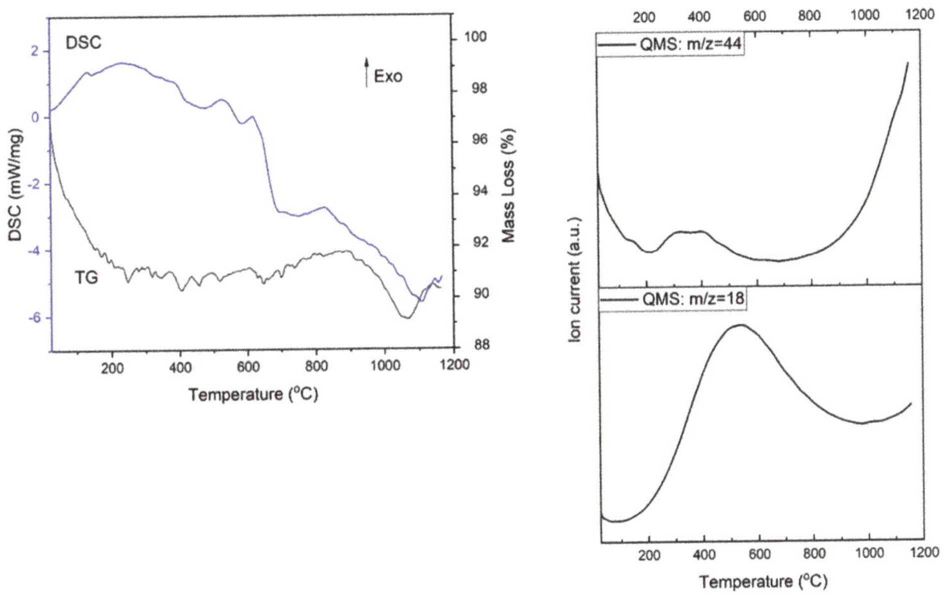

(**a**) (**b**)

Figure 8. EGA results for the pure filter, where: (**a**) presents DSC-TG signals, and (**b**) shows QMS registered signals.

Table 2. Volatile ions released from filters identified by QMS during thermal analysis process.

Registered Mass (*m*/*z*)	Chemical Formula	Registered Mass (*m*/*z*)	Chemical Formula
18	H_2O	45	CH_3CHOH, CH_2OCH_3, CH_2CH_2OH, COOH
22	doubly charged molecular ion, $CO_2{}^{2+}$.	46	NO_2
43	C_3H_7, $CH_3C{=}O$	58	C_2H_2S CH_3-CO-$CH_2{}^{+}H$
44	CO_2, CH_3CH-NH_2		

On the other hand, the examination of the filter from the Diesel engine shows similar behavior (Figure 10), but different gasses emitted were registered. From the room temperature up to approximately 600 °C, strong exothermic event on the DSC curve is visible, followed by less intense exothermic peek in the range from 600 to 700 °C. In this temperature range, the biggest weight loss of the sample was registered (~11 wt.%). This weight loss very likely is observed due to emission and decomposition of organic compounds collected on the filter. There is also some moisture contribution, as well (Table 2).

3.6. *Scanning Electron Microscope with Energy Dispersive Spectroscopy (SEM-EDS)*

More in-depth information was obtained using SEM-EDS that makes it possible to study the elementary analysis of the PM. Figure 11 shows morphology of the PM for diesel end petrol engines. It can be seen that particles deposited on the Diesel engine fibers are smaller, and their amount is lower (Figure 11a,b) than in case of petrol engine (Figure 11c,d). PM from the petrol engine showed more agglomerations. SEM observations in AsB mode showed different weight of the collected contaminations (heavier particles shines). EDS analysis described below reveals that shining objects are iron particles.

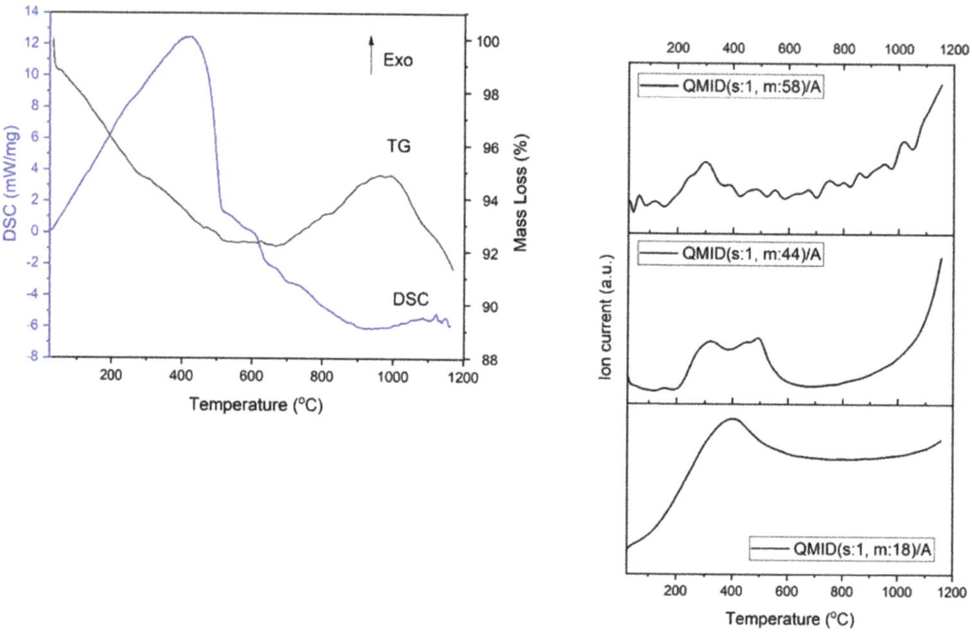

(**a**) (**b**)

Figure 9. EGA results for petrol engine, where (**a**) presents DSC-TG signals, and (**b**) shows QMS registered signals.

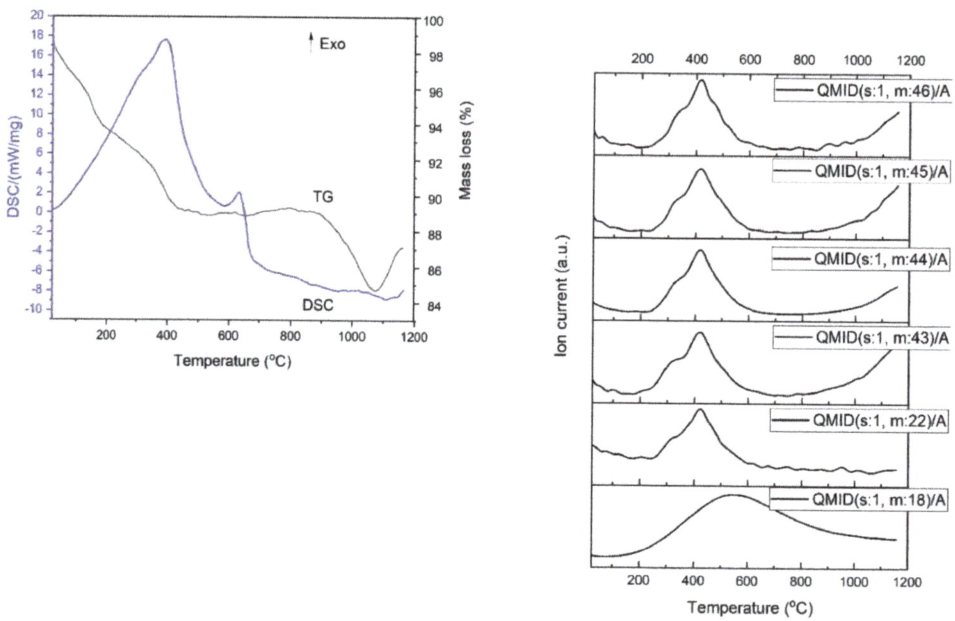

(**a**) (**b**)

Figure 10. EGA results for Diesel engine, where (**a**) presents DSC-TG signals, and (**b**) shows QMS registered signals.

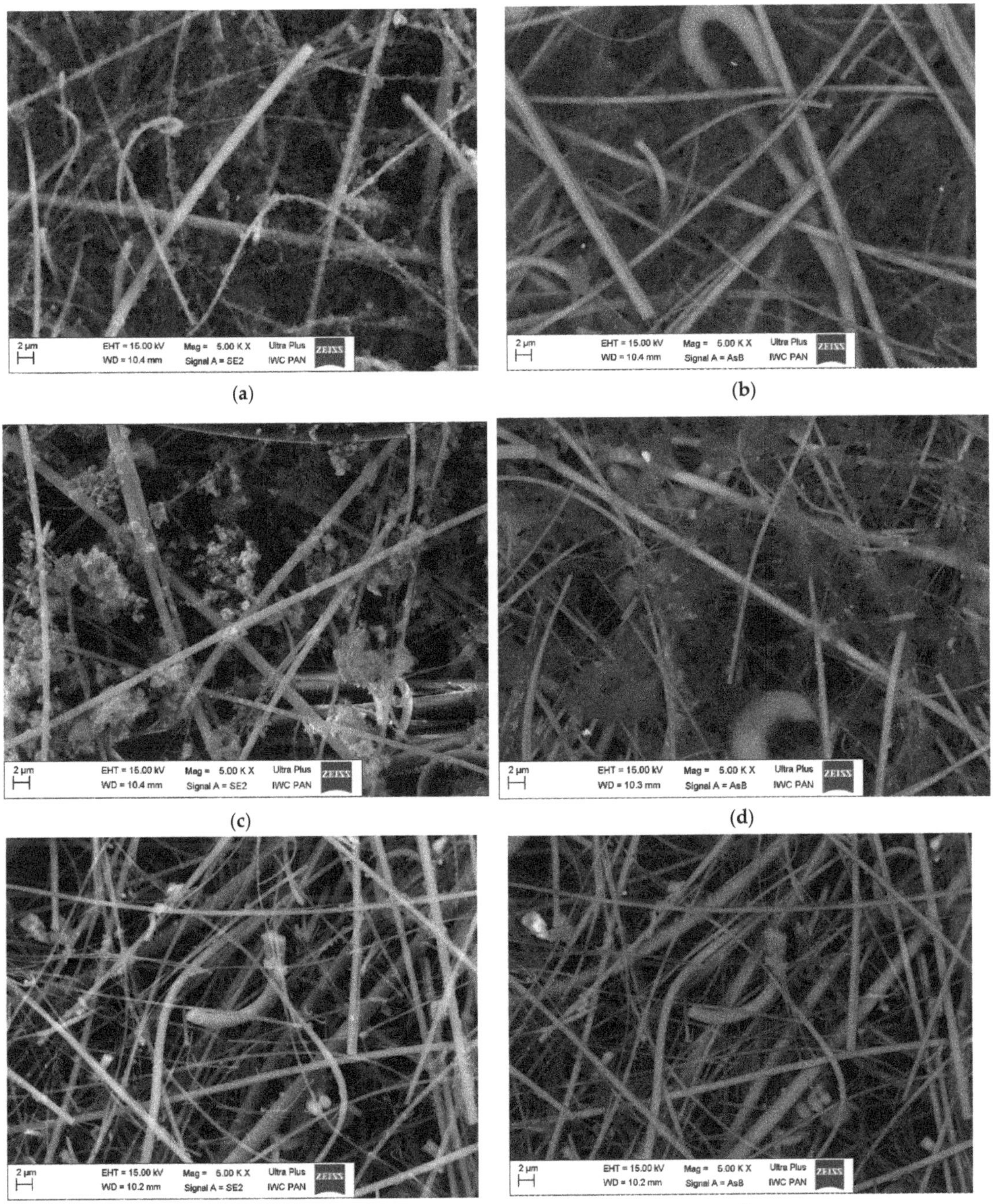

Figure 11. SEM images for Diesel engine in SE2 mode (**a**), and AsB mode (**b**); petrol engine (scooter) in SE2 mode (**c**), and AsB mode (**d**); pure filter in SE2 mode (**e**) and AsB mode (**f**).

The example of EDS area chosen from SEM micrographs for the Diesel and petrol (scooter) engine is shown in Figure 12.

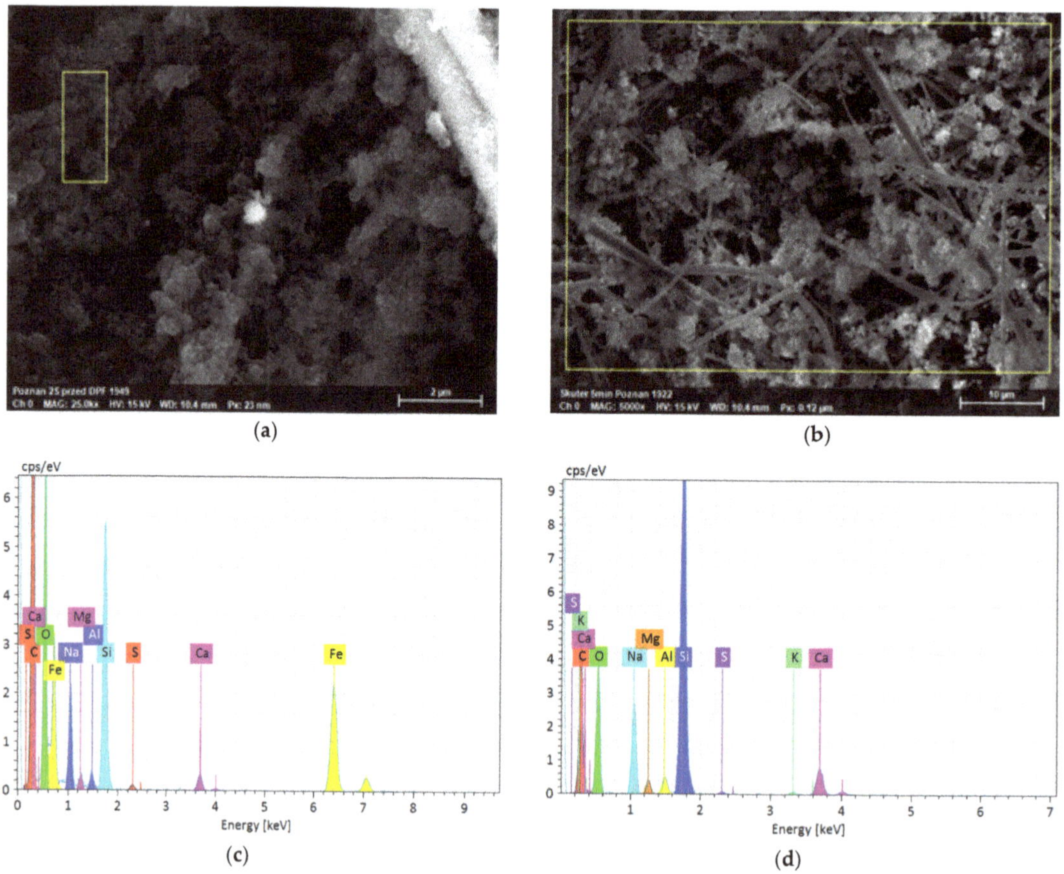

Figure 12. SEM micrographs for (**a**) Diesel engine and (**b**) petrol engine (scooter), showing area of EDS analysis. Elemental analysis for (**c**) Diesel and (**d**) scooter.

SEM-EDS analysis enabled to quantitatively assess the elemental analysis of PMs from the Diesel and petrol engine (Table 3). EDS analysis indicates that the main component of PMs from both engines is carbon (most probably from organic compounds originated from combustion process). The small amount on the pure filter is the effect of covering the filter by very thin layer of carbon (6 nm) before SEM-EDS imaging. The amount of oxygen is similar to the pure filter in the case of the Diesel engine, and it is significantly smaller for the petrol engine: this phenomenon can be the effect of covering filter fibers by carbon agglomerates in the case of the petrol engine and more distracted particles in the case of Diesel engine that do not stick the filter fibers. This phenomenon is also present in the case of Na, Ca, Si, Al, and Mg. Surprisingly, sulfur is present in the case of PMs from both engines, on a similar level that can come from some engine parts/oils. Only for PMs for Diesel engine was Fe noticed.

Table 3. Elemental Analysis of filters from Diesel and petrol engines.

Elemental Composition for Diesel Engine			Elemental Composition for Petrol Engine		Elemental Composition for Pure Filter	
Element	**Mass Norm. (%)**	**Abs. Error (1sigma) %**	**Mass Norm. (%)**	**Abs. Error (1sigma) %**	**Mass Norm. (%)**	**Abs.Error (1sigma) %**
Carbon	45.72	8.98	41.69	7.12	9.28	1.75
Oxygen	38.48	5.19	28.95	4.96	43.75	5.09
Sodium	3.65	0.39	5.83	0.48	7.46	5.09
Silicon	6.16	0.44	18.63	0.98	26.11	1.03
Calcium	0.84	0.09	2.86	0.15	1.43	0.09
Iron	18.02	0.91	Not detected	Not detected	Not detected	Not detected
Sulfur	0.16	0.05	0.15	0.05	Not detected	Not detected
Aluminum	0.38	0.07	0.90	0.09	2.88	0.16
Magnesium	0.38	0.08	0.77	0.09	Not detected	Not detected
Potassium	Not detected	Not detected	0.22	0.05	2.43	0.11

To compare these results with other available studies, it can be seen that, in research of Güney and Aladag [39], where gasoline-fueled vehicles were tested in terms of chemical analysis of particulate matter; in particular, SEM analyses show that PM is formed by solidification and agglomeration [39]. In this publication, the SEM analysis also showed that PM from gasoline engines shows more agglomerations, which was mentioned in point 3.6. In research of Güney and Aladag, the EDS analysis allowed for indicating 20 elements of PM: C, F, N, Na, O, Mg, Br, Si, Hg, S, P, Pb, Ca, Cr, Mn, Fe, Ni, Co, Cu, and Zn [39]. Some of the elements were also detected in SEM-EDS analysis in this paper, which is shown in Figure 12 and Table 3. Yang et al. [40] investigated the chemical composition of fine particulate matter emitted by gasoline and diesel vehicles, and this research showed that the most dominant components of PM2,5 for the Diesel and gasoline vehicle were carbonaceous species [40]. Research of Yang et al. also investigated that the top five metal elements in PM2,5 were Na, Ca, Fe, Zn, and Al, which also can be confirmed by studies of Cheung et al. [41] and Hao et al. [42].

4. Conclusions

The article presents the results of the analysis of the physical properties and chemical composition of particles emitted by a diesel and gasoline engine. The research focused not only on data obtained from particulate matter analyzer but also related to the chemical analysis of particulate matter from the tested engines, which allowed for comparing chemical composition and properties of particles emitted from diesel engine and petrol engine. There were three chemical analyses: Fourier Transform Infrared Spectroscopy, Scanning Electron Microscope with Energy Dispersive Spectroscopy, and Evolution Gas Analysis, which is a very extended analysis compared to the available studies. The diesel engine was mounted on an engine dynamometer; therefore, it was possible to analyze it in depth. In the case of the petrol engine, it was mounted on a scooter; therefore, a chassis dynamometer was used. The diesel engine emitted particles in the range of 50–120 nm. With the increase in the engine load, the specific emission of particulate matter increased. Increasing the load also resulted in an increase in the specific emission of the number and mass (particles per kilogram of fuel burnt). Additionally, the tests included the results before and after regeneration of the diesel particulate filter. In the case of a gasoline engine running without load, the emission of particles smaller than 30 nm was mainly observed. Increasing load of the gasoline engine resulted in an increase in both the concentration of particles and their diameter (average diameter to 90 nm).

Particles emitted from the Diesel and petrol engine differ in the size, number, and the chemical composition. PMs emitted from the Diesel engine are smaller, and more carbon black is emitted than in the case of petrol engine. It is a huge challenge to study the exact composition of PMs collected on the filters. Very accurate techniques were used in this paper, and it is possible to estimate the PMs chemical composition. EGA, SEM-EDS, and FTIR results show the differences of PMs chemical composition emitted from the Diesel and petrol engine. It can be concluded that PMs emitted from petrol engine agglomerate. The chemical composition is similar; however, for the Diesel engine, more carbon black is emitted. In case of both engines, sulfur compounds are observed.

On the basis of the conducted experimental studies, the following conclusions can be presented:

1. All adsorbed particulate matter collected from the Diesel and petrol engine contain carbon black. Carbon black shows absorption in the entire area from 4000 to 400 cm^{-1}, and, when using the ATR technique, the effect of carbon black becomes greater with deeper light penetration at the long wavelength end.
2. The largest amount of carbon black is present at particulate matter emitted from the Diesel engine, both working with and without load and during cold start. Slightly less carbon black is emitted in the case of an engine without load than with load.
3. There are no evidence signals from alkanes or aromatic compounds; however, the weak signals at ~2900 cm^{-1} at the range 1650–1400 cm^{-1} might be seen. It is larger for PM from the Diesel engine than the petrol engine.
4. It can be seen that particles deposited on the Diesel engine fibers are smaller, and their amount is lower, than in case of petrol engine. PM from the petrol engine show more agglomerations.
5. The main component of PM from both engines is carbon, which most likely comes from organic compounds derived from the combustion process.
6. The amount of oxygen is similar to the pure filter in the case of the Diesel engine, and it is significantly smaller for the petrol engine: this phenomenon can be the effect of covering filter fibers by carbon agglomerates in the case of the petrol engine and more distracted particles in the case of Diesel engine that do not stick the filter fibers. This phenomenon is also present in case of Na, Ca, Si, Al, and Mg.
7. Sulfur is present in particulate matter from both engines on the similar level.
8. Fe was noticed only for particulate matters from the Diesel engine.

Author Contributions: Conceptualization, R.J. and B.S.; methodology, R.J., B.S., I.K. and J.M.; software, R.J., B.S., I.K. and J.M.; validation, R.J. and B.S., formal analysis, R.J.; investigation, B.S.; resources, R.J.; data curation, R.J., B.S., I.K., J.M. and P.K.; writing—original draft preparation, R.J., B.S., I.K., J.M. and P.K., writing—review and editing, R.J.; visualization, B.S., I.K. and J.M.; supervision, R.J.; project administration, R.J.; funding acquisition, R.J. All authors have read and agreed to the published version of the manuscript.

Funding: The research subject was carried out with the use of equipment funded by the project CePT, reference: POIG.02.02.00-14-024/08, financed by the European Regional Development Fund within the Operational Programme "Innovative Economy" for 2007–2013.

Institutional Review Board Statement: Not applicable.

Informed Consent Statement: Not applicable.

Conflicts of Interest: The authors declare no conflict of interest.

References

1. Sicard, P.; Agathokleous, E.; De Marco, A.; Paoletti, E.; Calatayud, V. Urban population exposure to air pollution in Europe over the last decades. *Environ. Sci. Eur.* **2021**, *33*, 1–12. [CrossRef] [PubMed]
2. Lelieveld, J.; Evans, J.S.; Fnais, M.; Giannadaki, D.; Pozzer, A. The contribution of outdoor air pollution sources to premature mortality on a global scale. *Nature* **2015**, *525*, 367–371. [CrossRef] [PubMed]
3. OECD/European Union. *Health at a Glance: Europe 2020: State of Health in the EU Cycle*; OECD Publishing: Paris, France, 2020. [CrossRef]
4. World Health Organization. Air Pollution. Available online: https://www.who.int/health-topics/air-pollution#tab=tab_2 (accessed on 3 January 2021).
5. WHO. *Air Quality Guidelines for Particulate Matter, Ozone, Nitrogen Dioxide and Sulfur Dioxide*; WHO: Geneva, Switzerland, 2006.
6. Available online: https://whoairquality.shinyapps.io/AirQualityStandards/ (accessed on 29 June 2021).

7. Brook, R.D.; Rajagopalan, S.; Pope, C.A.; Brook, J.R.; Bhatnagar, A.; Diez-Roux, A.V.; Holguin, F.; Hong, Y.; Luepker, R.V.; Mittleman, M.A.; et al. Particulate Matter Air Pollution and Cardiovascular Disease: An update to the scientific statement from the american heart association. *Circulation* **2010**, *121*, 2331–2378. [CrossRef]
8. Jasiński, R.; Galant-Gołębiewska, M.; Nowak, M.; Ginter, M.; Kurzawska, P.; Kurtyka, K.; Maciejewska, M. Case Study of Pollution with Particulate Matter in Selected Locations of Polish Cities. *Energies* **2021**, *14*, 2529. [CrossRef]
9. Schikowski, T.; Vossoughi, M.; Vierkötter, A.; Schulte, T.; Teichert, T.; Sugiri, D.; Fehsel, K.; Tzivian, L.; Bae, I.-S.; Ranft, U.; et al. Association of air pollution with cognitive functions and its modification by APOE gene variants in elderly women. *Environ. Res.* **2015**, *142*, 10–16. [CrossRef]
10. Tonne, C.; Elbaz, A.; Beevers, S.; Singh-Manoux, A. Traffic-related Air Pollution in Relation to Cognitive Function in Older Adults. *Epidemiology* **2014**, *25*, 674–681. [CrossRef] [PubMed]
11. Ailshire, J.A.; Crimmins, E.M. Fine Particulate Matter Air Pollution and Cognitive Function Among Older US Adults. *Am. J. Epidemiol.* **2014**, *180*, 359–366. [CrossRef] [PubMed]
12. Gao, H.; Shi, J.; Cheng, H.; Zhang, Y. The impact of long- and short-term exposure to different ambient air pollutants on cognitive function in China. *Environ. Int.* **2021**, *151*, 106416. [CrossRef]
13. Pielecha, J.; Merkisz, J.; Markowski, J.; Jasiński, R. Analysis of Passenger Car Emission Factors in RDE Tests. *E3S Web Conf.* **2016**, *10*, 73. [CrossRef]
14. Guevara, M. Emissions of Primary Particulate Matter. *Issues Environ. Sci. Technol.* **2016**, 1–34. [CrossRef]
15. Reşitoğlu, İ.A.; Altinişik, K.; Keskin, A. The pollutant emissions from diesel-engine vehicles and exhaust aftertreatment systems. *Clean Technol. Environ. Policy* **2015**, *17*, 15–27. [CrossRef]
16. Report–Vehicles in Use, Europe 2019–ACEA–European Automobile Manufacturers' Association. 2021. Available online: https://www.acea.auto/publication/report-vehicles-in-use-europe-2019/ (accessed on 17 August 2021).
17. Dun, C.; Horton, G.; Kollamthodi, S. *Improvements to the Definition of Lifetime Mileage of Light Duty Vehicles*; Ricardo-AEA: London, UK, 2015.
18. Burtscher, H. Physical characterization of particulate emissions from diesel engines: A review. *J. Aerosol Sci.* **2005**, *36*, 896–932. [CrossRef]
19. Keskinen, J.; Rönkkö, T. Can Real-World Diesel Exhaust Particle Size Distribution be Reproduced in the Laboratory? A Critical Review Jorma Keskinen. *J. Air Waste Manag. Assoc.* **2010**, *60*, 1245–1255. [CrossRef] [PubMed]
20. Wróblewski, P.; Iskra, A. *Problems of Reducing Friction Losses of a Piston-Ring-Cylinder Configuration in a Combustion Piston Engine with an Increased Isochoric Pressure Gain*; SAE Technical Paper Series, SAE Technical Paper 2020-01-2227; SAE International: Warrendale, PA, USA, 2020. [CrossRef]
21. R'Mili, B.; Boréave, A.; Meme, A.; Vernoux, P.; Leblanc, M.; Noël, L.; Raux, S.; D'Anna, B. Physico-Chemical Characterization of Fine and Ultrafine Particles Emitted during Diesel Particulate Filter Active Regeneration of Euro5 Diesel Vehicles. *Environ. Sci. Technol.* **2018**, *52*, 3312–3319. [CrossRef]
22. Staps, I.; Ligterink, N. *Diesel Particle Filters*; TNO: The Hague, The Netherlands, 2018. [CrossRef]
23. Steininger, N. Particle number emission limits for Euro 6 positive ignition vehicles (PI). In Proceedings of the 15 ETH Conference on Combustion Generated Nano-Particles, Zurich, Switzerland, 27–29 June 2011.
24. Diesel Particulate Filters-GOV.UK. Retrieved 17 August 2021. Available online: https://www.gov.uk/government/publications/diesel-particulate-filters-guidance-note/diesel-particulate-filters (accessed on 17 August 2021).
25. Green-Zones.eu.—Green-Zones.eu. Retrieved 17 August 2021. Available online: https://www.green-zones.eu/en/ (accessed on 17 August 2021).
26. Plötz, P.; Axsen, J.; Funke, S.A.; Gnann, T. Designing car bans for sustainable transportation. *Nat. Sustain.* **2019**, *2*, 534–536. [CrossRef]
27. Chang, A.S.F.; Kalawsky, R.S. European transport sector interventions for smart city. In Proceedings of the 2017 7th International Conference on Power Electronics Systems and Applications—Smart Mobility, Power Transfer & Security (PESA), Hong Kong, China, 12–14 December 2017; IEEE: New York, NY, USA, 2017; Volume 5, pp. 1–6. [CrossRef]
28. National Research Council. *Cost, Effectiveness, and Deployment of Fuel Economy Technologies for Light-Duty Vehicles*; National Academies Press: Cambridge, MA, USA, 2015.
29. Conway, G.; Joshi, A.; Leach, F.; García, A.; Senecal, P.K. A review of current and future powertrain technologies and trends in 2020. *Transp. Eng.* **2021**, *5*, 100080. [CrossRef]
30. Macioszek, E.; Sierpiński, G. Charging Stations for Electric Vehicles—Current Situation in Poland. *Commun. Comput. Inf. Sci.* **2020**, *13*, 124–137. [CrossRef]
31. Wróblewski, P.; Drożdż, W.; Lewicki, W.; Dowejko, J. Total Cost of Ownership and Its Potential Consequences for the Development of the Hydrogen Fuel Cell Powered Vehicle Market in Poland. *Energies* **2021**, *14*, 2131. [CrossRef]
32. Li, Q.; Zhu, Q.; Xu, M.; Zhao, Y.; Narayan, K.; Liu, Y. Estimating the Impact of COVID-19 on the $PM_{2.5}$ Levels in China with a Satellite-Driven Machine Learning Model. *Remote Sens.* **2021**, *13*, 1351. [CrossRef]
33. Rogula-Kozłowska, W.; Klejnowski, K.; Rogula-Kopiec, P.; Ośródka, L.; Krajny, E.; Błaszczak, B.; Mathews, B. Spatial and seasonal variability of the mass concentration and chemical composition of PM2.5 in Poland. *Air Qual. Atmos. Health* **2014**, *7*, 41–58. [CrossRef]

34. Popovicheva, O.B.; Kireeva, E.D.; Shonija, N.K.; Vojtisek-Lom, M.; Schwarz, J. FTIR analysis of surface functionalities on particulate matter produced by off-road diesel engines operating on diesel and biofuel. *Environ. Sci. Pollut. Res.* **2014**, *22*, 4534–4544. [CrossRef]
35. Mico, S.; Tsaousi, E.; Deda, A.; Pomonis, P. Characterization of airborne particles and source identification using SEM/EDS. *Eur. Chem. Bull.* **2015**, *4*, 224–229. [CrossRef]
36. Genga, A.; Baglivi, F.; Siciliano, M.; Siciliano, T.; Aiello, D.; Tortorella, C. Chemical and Morphological Study of Particulate Matter Analysed by Sem-Eds. In Proceedings of the 4th Imeko TC19 Symposium on Environmental Instrumentation and Measurements Protecting Environment, Climate Changes and Pollution Control, Lecce, Italy, 3–4 June 2013; pp. 43–46.
37. Kurzawska, P.; Jasiński, R. Overview of Sustainable Aviation Fuels with Emission Characteristic and Particles Emission of the Turbine Engine Fueled ATJ Blends with Different Percentages of ATJ Fuel. *Energies* **2021**, *14*, 1858. [CrossRef]
38. Merkisz, J. On-road exhaust emission testing. *Combust. Engines* **2011**, *146*, 3–15. [CrossRef]
39. Güney, B.; Aladağ, A. Microstructural characterization of particulate matter from gasoline-fuelled vehicle emissions. *J. Eng. Res. Rep.* **2020**, *16*, 29–39. [CrossRef]
40. Yang, H.H.; Dhital, N.B.; Wang, L.C.; Hsieh, Y.S.; Lee, K.T.; Hsu, Y.T.; Huang, S.C. Chemical characterization of fine particulate matter in gasoline and diesel vehicle exhaust. *Aerosol Air Qual. Res.* **2019**, *19*, 1349–1449. [CrossRef]
41. Cheung, K.L.; Ntziachristos, L.; Tzamkiozis, T.; Schauer, J.J.; Samaras, Z.; Moore, K.F.; Sioutas, C. Emissions of particulate trace elements, metals and organic species from gasoline, diesel, and biodiesel passenger vehicles and their relation to oxidative potential. *Aerosol Sci. Technol.* **2010**, *44*, 500–513. [CrossRef]
42. Hao, Y.; Gao, C.; Deng, S.; Yuan, M.; Song, W.; Lu, Z.; Qiu, Z. Chemical characterisation of PM2.5 emitted from motor vehicles powered by diesel, gasoline, natural gas and methanol fuel. *Sci. Total Environ.* **2019**, *674*, 128–139. [CrossRef] [PubMed]

energies

MDPI

Article

Pollutant Emissions during Oxy-Fuel Combustion of Biomass in a Bench Scale CFB Combustor

Monika Kosowska-Golachowska [1,*], Adam Luckos [2] and Agnieszka Kijo-Kleczkowska [1]

1 Department of Thermal Machinery, Czestochowa University of Technology, Armii Krajowej 21, 42-201 Czestochowa, Poland; a.kijo-kleczkowska@pcz.pl
2 School of Chemical and Metallurgical Engineering, University of the Witwatersrand, Braamfontein Campus East, Johannesburg 2050, South Africa; adam.luckos@wits.ac.za
* Correspondence: m.kosowska-golachowska@pcz.pl

Abstract: Nowadays oxy-fuel combustion of coal and biomass is the most promising option for the reduction of CO_2 emissions from power plants. In this paper, emissions of NO_x (NO, NO_2, N_2O and their precursors, such as NH_3 and HCN), SO_2 and CO during conventional and oxy-fuel combustion of three kinds of biomass (agro, woody and energy crop) and a reference coal are presented and discussed. Combustion tests were conducted at 850 °C in the laboratory-scale circulating fluidized bed (CFB) reactor in air and O_2/CO_2 atmospheres. A FTIR spectrometer was used to measure instantaneous concentrations of all pollutants in the flue gas. Emissions of SO_2, N_2O and CO for the combustion of biomass in all atmospheres were lower than those for the combustion of reference coal. It was found that oxidation of nitrogen species released with volatile matter was responsible for high emissions of NO_x during combustion of biomass fuels in air and mixtures of O_2 and CO_2. The lowest NO emissions for tested fuels were detected in oxy-21 atmosphere (21% O_2/70% CO_2). Oxy-combustion of biomass in O_2/CO_2 mixtures at 30% and 40% O_2 caused a decrease in emissions of N_2O and CO while NO and SO_2 emissions increased. The results of this study show that the tested biomass fuels are ideal renewable energy resources both in conventional and oxy-fuel conditions with a minor potential for environmental pollution.

Keywords: oxy-combustion; circulating fluidized bed; wheat straw; *Salix viminalis*; Scots pine; renewable energy sources; NOx; SO_2; CO; CO_2

Citation: Kosowska-Golachowska, M.; Luckos, A.; Kijo-Kleczkowska, A. Pollutant Emissions during Oxy-Fuel Combustion of Biomass in a Bench Scale CFB Combustor. *Energies* **2022**, *15*, 706. https://doi.org/10.3390/en15030706

Academic Editor: Maria Founti

Received: 5 December 2021
Accepted: 14 January 2022
Published: 19 January 2022

1. Introduction

Utilization of fossil fuels, mainly coal, for power generation is associated with emissions of large quantities of carbon dioxide—a greenhouse gas responsible for global warming and climate change. As the world will still depend on coal for several decades to come, it is necessary to develop technologies such as carbon capture and storage (CCS) to significantly reduce emissions of CO_2 and avoid negative consequences of global warming. Among CCS technologies, oxy-fuel combustion emerges as a technically feasible and cost-effective option for the reduction of CO_2 emissions from coal-fired power plants [1–3]. When combined with co-firing of biomass—a renewable carbon-neutral fuel—it can offer negative CO_2 emissions [4–6].

In an oxy-fuel combustion system, fuel is burned in a mixture of oxygen and recycled flue gas. As molecular nitrogen is eliminated from the oxidizing medium, the flue gas consists mainly of CO_2 and water vapor. Partial recirculation of flue gas is necessary to control the combustion temperature in the boiler. Substituting recycled flue gas for N_2 in the oxidizing medium leads to a smaller exhaust gas stream, higher boiler efficiency and lower NO_x emissions [3].

Nowadays, pulverized coal (PC) combustion is the dominant technology of heat and power generation in utility and industrial sectors. However, circulating fluidized-bed (CFB) combustion gains popularity, particularly for the utilization of low quality, high-ash coals, and discards. Today, high-efficiency, utility CFB units with supercritical steam parameters and electric outputs up to 600 MW are available commercially. Fluidized bed combustion is also considered the best available co-firing technology [4,7–9]. Co-firing of coal and biomass in an oxy-CFB boiler has been demonstrated at the 30 MW_{th} scale at Fundación Ciudad de la Energía (CIUDEN) in Spain [10,11].

In the case of oxy-combustion, CFB boilers have a few advantages over PC counterparts. The combustion temperature in oxy-CFB systems can be controlled by cooling (in an external heat exchanger) and recycling a part of circulating solid particles. This reduces the stream of recirculated flue gas and makes an oxy-CFB boiler smaller, more efficient, and cheaper. Due to lower combustion temperature, oxy-CFB units can meet NO_x and SO_x emission limits without additional de-NO_x and de-SO_x systems. Moreover, the oxy-CFB technology does not require sophisticated burner and fuel feeding systems.

A significant progress has been made in the case of oxy-combustion in PC boilers. However, to date, only a few studies have been published on combustion or co-combustion of biomass and coal in oxy-fuel fluidized-bed combustors. Some few among them are devoted to the formation and emissions of pollutants such as NO_x, SO_2 and CO.

Carbon dioxide storage/sequestration requires a gas stream with 95% or more CO_2 purity. To achieve that a CO_2 processing unit (CPU) is employed for drying, cleaning and compression of the exhaust gas. The presence of impurities affects the design and energy consumption in the CPU. The impurities of interest are O_2, N_2 and Ar supplied by the ASU and related to the purity of oxygen produced and excess O_2 used in the oxy-combustion system. They also include NO_x, SO_2 and other N- and S-containing gases such as HCN, NH_3, COS and H_2S. Therefore, the knowledge on pollutant formation and their concentration in the flue gas is of crucial importance for the proper design of the CPU and assessment of its performance and cost.

In this study, combustion tests of three kinds of biomass (agricultural, woody and energy crop) and a reference coal were carried out in the laboratory-scale CFB reactor at 850 °C. Concentrations of NO_x (NO, NO_2, N_2O and their precursors, such as NH_3 and HCN), SO_2 and CO were measured online during conventional and oxy-fuel combustion. The main objective was to assess the influence of oxidizing atmosphere and composition of tested biomass on the formation mechanism and emissions of pollutants. A comparison with emissions from bituminous coal can be useful to determine emission levels in the case of co-firing coal and biomass.

The study revealed that oxidation of nitrogen species released with volatile matter was responsible for high emissions of NO_x during combustion of biomass fuels in air and mixtures of O_2 and CO_2. Emissions of SO_2, N_2O and CO for the combustion of biomass in all atmospheres were lower than those for the combustion of reference coal. Oxy-combustion of biomass in O_2/CO_2 mixtures at 30% and 40% O_2 caused a decrease in emissions of N_2O and CO while NO and SO_2 emissions increased.

2. Literature Review

This short literature review is devoted to pollutant emissions from oxy-combustion of biomass fuels in fluidized-bed systems. The only exception is the study by Gao and co-workers [12] that was carried out in a TG-MS system with algae *Chlorella vulgaris*, a promising biofuel for "green" electricity generation.

Tan and co-workers [13] conducted oxy-combustion co-firing tests with different coals and wood pellets in the CanmetENERGY 800 kW_{th} CFB combustor (406 mm ID). The fraction of cofired biomass was in the range 20–50%wt. The combustion characteristics, pollutant and trace metals emissions were measured at O_2 concentration in the combustion gas between 24 and 25%vol. The authors concluded that the addition of wood pellets did not have a significant influence on combustion conditions. Emissions of NO were in a narrow range of 14–20 ng/J of heat input. All tests were carried out with the addition of limestone at the Ca/S ratio of 3. Emissions of SO_2 varied from 35 to 95 ng/J of heat input. Concentrations of CO were stable and below 200 ppm when the O_2 concentration in the flue gas exceeded 3.5%. In all tests, the NO and SO_2 emission rates were below limits set by Canada and the UE.

Duan and co-workers [5] investigated NO emissions during oxy-fuel co-firing of biomass (rice husk, wood chips and dry wood flour) and bituminous coal in a 10 kW_{th} CFB combustor (65 mm ID, 2.3 m height). The ratio of biomass in the fuel blend was in the range 0–100%wt. All tests were carried out at the same conditions—bed temperature 850 °C, excess oxygen 5% and primary O_2 fraction 0.7. In the case of single fuel combustion, NO emissions were in the range 56–195 $mg/m^3{}_n$ and they were greater for biomass fuels with the highest value observed for rice husk and the lowest for wood chips. Oxy-fuel combustion produced less NO than combustion in air. The conversion of fuel-N to NO increased with increasing H/N mass ratio in tested fuel. In the case of co-firing biomass and coal, NO emissions increased with increasing fraction of biomass in the fuel blend for all biomass fuel tested under air and oxy-fuel conditions.

Lupiáñez and co-workers [14] studied oxy-combustion of anthracite and corn stover blends in a laboratory-scale bubbling fluidized bed combustor (203 mm ID, 2.5 m height) with thermal input around 30 kW. Combustion tests at bed temperatures 850–900 °C in air and 30/70% mixtures of O_2 and CO_2 were carried out with 80/20 blends (on energy basis) of coal and biomass. In two tests, limestone was added for the in situ SO_2 capture at the Ca/S molar ratio of 2.5. Measurements of the flue gas composition revealed that emissions of SO_2 (corrected to 6% O_2 and normalized to mg/MJ) were in the range from 105 (oxy-combustion at 870 °C) to 906 mg/MJ (air combustion at 860 °C). These emissions were related to the chlorine content in corn stover; the higher the Cl content in corn stover the lower the SO_2 emission. The desulphurization efficiency exceeding 80% was achieved with limestone addition. Emissions of NO were not influence by the Cl in corn stover and they were in the range from 68–172 mg/MJ. According to the authors, the expected reduction in NO emissions during oxy-firing was not observed, owing to the higher excess of O_2. Finally, it was concluded that the influence of coal/biomass ratio on NO emissions was negligible. The catalytic activity of limestone and higher O_2 excess were responsible for the larger conversion of fuel-N to NO during oxy-combustion tests.

In the subsequent studies [15,16] Lupiáñez and co-workers investigated the influence of limestone on gaseous emissions during oxy-combustion of lignite and corn stover in the fluidized-bed combustor described in the previous paper. The author founded that the SO_2 capture was influence by the limestone type and fragmentation of its particles. Emissions of SO_2 decreases with the Ca/S ratio and bed temperature. On the other hand, emissions of NO increased with Ca/S ratio and the presence of calcined limestone in the bed. They were more influenced by the excess of O_2 and CO concentration in the bed than by the biomass share in the blend or chlorine content in the fuel.

Pu and co-workers [17] studied the influence of oxygen concentration in the oxidizing medium (O_2/CO_2 mixture with O_2 concentrations in the range 21–40%) on NO emissions in a lab-scale bubbling fluidized-bed combustor at 850–950 °C. Combustion tests were carried out with 2 g samples of anthracite and anthracite blended with 10–30%wt. of pine powder. At 900 °C, the conversion rate of fuel-N to NO increased with increasing O_2 concentration from 40.4% to 42.6% in the case of anthracite and from 35.1% to 41.2% in the case of 20% pine powder blend. Tests with 35% O_2/65% CO_2 at different temperatures revealed a slight increase in NO emissions with increasing temperature in the case of anthracite. The opposite trend was observed for the blend containing 20% pine powder. The total NO emission was reduced from 12.48 mg at 850 °C to 11.46 mg at 950 °C which corresponded to a 3.4% decrease in the fuel-N conversion to NO. The influence of pine powder fraction in the blend was studied at 950 °C with 35% O_2 in the O_2/CO_2 mixture. It was found that the addition of pine powder reduced the NO emission and fuel-N conversion to NO. The fuel-N conversion decreased from 42.9% in the case of anthracite to 37.8% in the case of blend containing 30% pine powder.

Wang and co-workers [18] investigated nitrogenous gas (NO, N_2O, HCN) emissions from co-combustion of coal and biomass (corn straw and wheat straw) under oxy-fuel conditions with 50% oxygen concentration in the oxidizing medium. Tests were carried out in a 100 kW CFB combustor with blends containing 30% biomass and in atmospheres containing O_2/CO_2 and O_2/recycled flue gas mixtures (RFG) at temperatures 800–900 °C. As expected, the NO and N_2O emissions increased with increasing excess oxygen in the oxidizing medium. Tests with different fractions of corn straw in the fuel (10, 20 and 30%) revealed an increase in emission factors of NO, N_2O, HCN by 241, 238 and 249%, respectively. The results of tests with coal and wheat straw at 800 and 900 °C showed that the NO emission factor increased slightly with increasing temperature. On the other hand, the emission factors of N_2O and HCN were lower at 900 °C than those at 800 °C.

Sher and co-workers [19] studied the influence of the oxidizing medium (air, $O_2/N_2/CO_2$ mixtures) on gaseous emissions and temperature profile during combustion of miscanthus, straw pellets and wood pellets in a 20 kW_{th} fluidized-bed combustor. The authors observed a significant decrease in CO emissions for all three fuels when O_2 concentration in the oxy-fuel medium exceeded 25%. Emissions of NO decreased with increasing O_2 concentration in the oxy-fuel medium and at 30% O_2 they were like those for combustion in air. It was found that NO concentrations in the flue gas are related to the nitrogen content in the fuel tested.

Varol and co-workers [8] investigated oxy-fuel combustion of high-sulfur lignite and wood pellet blends (up to 60%) in a laboratory-scale (100 mm ID, 5.1 m height) CFB combustor. The main objective of their study was to determine the effect of biomass share on NO_x, SO_2 and CO emissions. The results showed that increasing biomass share in the fuel blend had a negligible influence on NO_x emissions. Emissions of CO decreased slightly with an increase in biomass share. Concentrations of SO_2 in the flue gas were related to the sulfur content in the fuel and they decreased with increasing biomass share in the blend.

Nguyen and co-workers [20] studied SO_2, NO and CO emissions during co-combustion of lignite and wood pellets (50–100%wt. share in the fuel) in an oxy-fuel 100 kW_{th} test facility. All tests were carried out with 25% oxygen in the oxidizing medium. An increase in biomass share caused a decrease in NO, SO_2 and CO concentrations from 19.2 mg/MJ (corrected to 6% O_2 in the flue gas) to 16.1 mg/MJ, 92.8 mg/MJ to 25.0 mg/MJ, and 7.5 mg/MJ to 5.5 mg/MJ, respectively. The authors concluded that oxy-combustion of pure biomass can lead to negative CO_2 emissions of, approximately, −647 g/kW_{th}.

Gao and co-workers [12] investigated pollutant formation during air and oxy-fuel combustion of microalgae *Chlorella vulgaris* and its blends with Chinese lignite in a TG-MS (thermogravimetry coupled with mass spectrometry) system. The presence of microalgae in the fuel blend during combustion in air resulted in lower emissions of CO_2, CO and NO_2, but in enhanced formation of NO, COS and SO_2. Similar trend was observed in the case of oxy-fuel combustion of microalgae/lignite blends.

The main conclusions of the conducted review are summarized in Table 1.

Table 1. A summary of the literature review.

Reference	Test Facility and Conditions	Fuels Tested	Composition of Oxidizing Medium	Emissions Reported	Remarks
Tan et al. [13]	CFB, 800 kW_{th}, ~900 °C, limestone, Ca/S = 3	Wood pellets, different coals, fraction of biomass in blends 20–50%wt.	O_2/CO_2 mixtures, 24–25% O_2, recycled flue gas	NO_x, SO_2, CO, CO_2, O_2	The addition of biomass did not have a significant influence on combustion conditions. Emissions of NO, SO_2 and CO below the EU limits (200 ppm).

Table 1. *Cont.*

Reference	Test Facility and Conditions	Fuels Tested	Composition of Oxidizing Medium	Emissions Reported	Remarks
Duan et al. [5]	CFB, 10 kW_{th}, 850 ± 10 °C	Rice husk, wood chips, dry wood flour, bituminous coal, biomass/coal ratio 0–100%wt.	Air, 70% O_2/30% CO_2 mixture	NO, NO_2, CO, CO_2, O_2, SO_2	For single fuel combustion, emissions of NO were higher for biomass (the highest for rice husk) than for coal. Oxy- combustion produced less NO than combustion in air. NO emissions increased with increasing fraction of biomass in the fuel blend for all biomass fuels in air and oxy-fuel conditions.
Lupiáñez et al. [14]	BFB *, 50 kW_{th}, 850–900 °C, limestone, Ca/S = 2.5	Corn stover, anthracite, biomass fraction in blend 0.2 (on energy basis)	Air, 30% O_2/70% CO_2 mixture	NO, CO, CO_2, O_2, SO_2	No reduction in NO emissions during oxy-firing was observed. Influence of coal/biomass ratio on NO emissions was negligible. Emissions of SO_2 higher in air than in oxy-combustion.
Lupiáñez et al. [15,16]	BFB *, 50 kW_{th}, 850–925 °C, limestone, Ca/S ratio: 2.5, 4, 6	Corn stover, lignite, biomass fraction in blend 0.1 and 0.2	Air, 35% O_2/65% CO_2, 30% O_2/70% CO_2	NO, SO_2, CO, CO_2, O_2, HCl	Emissions of NO increased with increasing Ca/S ratio. They were more influenced by the excess O_2 and concentration of CO than by the biomass share in the fuel blend.
Pu et al. [17]	BFB *, 850–950 °C	Pine powder, anthracite, biomass fraction in blend 0–30%wt.	21–40% O_2, CO_2	NO, NO_2, CO, CO_2, O_2, SO_2	Concentration of NO increased with increasing initial concentration of O_2 in the oxidizing medium. Addition of pine powder reduced the NO emission in the test at 950 °C with 35% O_2 in the O_2/CO_2 mixture.
Wang et al. [18]	CFB, 10 kW_{th}, 800–900 °C	Corn straw, wheat straw, coal, 30% biomass in blend	50% O_2/50% CO_2, 50% O_2/50% recycled flue gas	NO, N_2O, CO, CO_2, O_2, HCN	NO and N_2O emissions increased with increasing excess O_2. An increase in the fraction of corn straw in the fuel blend caused an increase in emission factors of NO, N_2O, and HCN.
Sher et al. [19]	BFB *, 20 kW_{th}, 800 °C	Miscanthus, straw pellets, wood pellets	Air, O_2/N_2/CO_2 mixtures: 21/22/57%, 24/25/52%, 27/28/45%	NO_x, CO, CO_2, O_2	Emissions of CO and NO decreased with increasing concentration of O_2 in the oxy-fuel medium. At 30% O_2, emissions of NO were like those for combustion in air.

Table 1. *Cont.*

Reference	Test Facility and Conditions	Fuels Tested	Composition of Oxidizing Medium	Emissions Reported	Remarks
Varol et al. [8]	CFB, 850 and 915 °C, limestone, Ca/S = 2	Wood pellets, high-sulphur lignite, fraction of biomass in blend up to 60%	25, 30% O_2, CO_2	NO_x, SO_2, CO, CO_2	Increasing biomass share in the fuel blend had a negligible influence on NO_x emissions. Emissions of CO and SO_2 decreased with increasing fraction of biomass in the fuel blend.
Nguyen et al. [20]	CFB, 100 kW_{th}	Wood pellets, lignite, fraction of biomass in blend 50–100%wt.	21–29% O_2, CO_2	NO, SO_2, CO	An increase in biomass share caused a decrease in NO, SO_2 and CO concentrations. Oxy-combustion of pure biomass can produce negative CO_2 emissions of, approximately, –647 g/kW_{th}.
Gao et al. [12]	TG-MS **, non-isothermal tests, 20–800 °C, heating rates 5–15 K/min	Microalgae *Chlorella vulgaris*, lignite, 50/50 blends	21% O_2/79% N_2, 21% O_2/79% CO_2	CO, CO_2, CO_S, NO, NO_2, SO_2	The presence of microalgae in the fuel blend during combustion in air resulted in lower emissions of CO_2, CO and NO_2 but in enhanced formation of NO, COS and SO_2. Similar trend was observed during oxy-combustion.

* BFB—bubbling fluidized-bed combustor, ** TG-MS—thermogravimetric analyzer with mass spectrometer.

3. Materials and Methods

3.1. Fuel Tested

Wheat straw (agricultural biomass), willow (*Salix viminalis*, energy crop biomass) and Scots pine (woody biomass) were used as primary fuels, and a Polish bituminous coal, combusted in CFB boilers, was chosen as the reference fuel.

The moisture content in biomass fuels was determined with the use of the oven-drying method, based on the PN-EN ISO 18134. The ash content of the biomass samples was determined in accordance with PN-EN ISO 18122, whereas PN-EN ISO 18123 standard was used to assess the volatile matter content. Elemental analyzers Truespec CHN Leco and SC-144DR Leco were used to determine the hydrogen, carbon, sulfur and nitrogen contents. The proximate and ultimate analyses and higher heating value (HHV) of the tested fuels are presented in Table 2.

Table 2. Proximate and ultimate and higher heating value of the tested biomass and coal.

	Wheat Straw	*Salix viminalis*	Scots Pine	Bituminous Coal [21]
Proximate analysis (air-dry basis), wt.%				
Moisture (M)	8.4	6.9	7.0	8.7
Ash yield (A)	6.1	1.4	0.6	18.9
Volatile matter (VM)	68.3	76.3	76.8	26.8
Fixed carbon (FC), (by difference)	17.2	15.4	15.6	45.6
Higher heating value (HHV), MJ/kg	17.84	18.20	18.90	22.75
Ultimate analysis (dry, ash-free basis), wt.%				
Carbon (C)	50.20	49.60	50.90	73.30
Hydrogen (H)	5.80	6.00	5.70	4.30
Sulphur (S)	0.08	0.03	0.01	2.30
Nitrogen (N)	0.80	0.30	0.10	1.10
Chlorine (Cl)	0.15	0.01	0.01	0.70
Oxygen (O), (by difference)	42.97	44.06	43.28	18.30

3.2. Experimental Setup and Procedure

Conventional and oxy-fuel combustion experiments were carried out in a 12-kW, electrically heated lab-scale CFB combustor, which has been described in detail in references [21–23]. Here, only the experimental conditions are briefly mentioned. All experiments were carried out at temperature 850 °C in air (base case) and mixtures of O_2/CO_2 with O_2 concentrations in the range from 21% to 40%vol. The names given to the different oxy-fuel atmospheres, oxy-21, oxy-30 and oxy-40, are related to the O_2 concentration in the oxy-fuel mixture fed to the combustor, i.e., 21, 30 and 40%vol., respectively.

Silica sand (particles below 400 μm) to a mass of 0.3 kg constituted the fluidized bed. The gases (O_2 and CO_2) to make up gas mixtures are supplied from cylinders to a mixer and then transferred via a preheater directly into the combustion chamber. Flow rates of gases are controlled by valves and measured by rotameters. During air and oxy-combustion experiments, the superficial gas velocity was kept at a constant level of about 5 m/s.

A small sample of fuel (0.5 g) was fed into the combustion chamber (riser) for each test. Sorbent was not used to capture SO_2 during all tests. The biomass and coal samples were in the form of spherical pellets. The detail method of pellet making is explained by Kosowska-Golachowska et al. [24] and Kijo-Kleczkowska et al. [25]. Concentrations of NO, NO_2, N_2O, SO_2, CO and other compounds (HCN and NH_3) in the flue were measured online (with frequency of 1 Hz) by a FTIR spectrometer (Gasmet DX-4000). The maximum error of these measurements was less than 2%. The time period for each test was set 3 s. Each test was repeated minimum three times to guarantee a relative standard deviation of less than 5%. This methodology has been successfully used in our previously studies on pollutant emissions during combustion of sewage sludge [21] or lignite and bituminous coal [22].

4. Results and Discussion

Biomass combustion both in air and oxy-fuel atmospheres proceeds through several stages, namely drying, devolatilization, ignition and volatile matter combustion, char combustion and agglomeration of ash [7,22,23]. A detail report on all these stages, temperature profiles and visualization during oxy-combustion of wheat straw and *Salix viminalis* can be found in [22,23], respectively.

Volatile matter and oxygen contents in the biomass samples are significantly higher than those in the bituminous coal whereas the ash yield and HHV are lower than those in coal (see Table 1). The contents of S, N and Cl in biomass are lower than those in coal, which indicates that the tested biomass fuels are ideal renewable energy resources with a minor potential for environmental pollution.

In this section, emissions of NO_x (NO, NO_2, N_2O and their precursors, such as NH_3 and HCN), SO_2 and CO during conventional and oxy-fuel combustion of three kinds of biomass and a reference coal are presented and discussed.

4.1. Conventional Combustion

4.1.1. NO, NO_2 and N_2O Emissions

Nitrogen oxides (NO_x), such as NO, NO_2 and N_2O, are main contributor to air pollution [26]. The mechanism of NO_x formation in conventional air-fired combustion includes thermal, prompt and fuel-N NO_x. Thermal and prompt NO_x formation occurs at high temperatures, usually above 1400 °C, that are typical for pulverized fuel combustion. In fluidized-bed combustion, contributions of thermal and prompt mechanisms to the total NO_x emissions are negligible.

Fuel nitrogen oxides are formed through oxidation of N-containing species in the fuel (fuel-N), which are released during devolatilization (volatile-N) and char oxidation (char-N) stages. At typical fluidized-bed combustion temperatures (below 900 °C), fuel-N is the dominant source for NO_x formation. Although char is an important intermediate for NO_x formation, the behavior of volatile-N is a dominating factor in the case of burning fuels with high-volatile matter contents such as biomass [27].

Figure 1 shows the time-resolved emissions of NO and N_2O during air-CFB combustion of all tested fuels. Nitrogen oxide was the dominant N-containing pollutant and no NO_2 was detected during tests in air. This finding is consistent with the results obtained by Liu et al. [28]. The highest NO concentrations (Figure 1a) were detected during the volatile matter combustion. The highest NO concentrations were approximately 75 ppm and 60 ppm for wheat straw and *Salix viminalis*, respectively. The lowest NO emissions were measured for the reference coal despite its highest nitrogen content. This observation can be attributed to the lower volatile matter content in coal and lower temperature during the volatile matter combustion compared to biomass fuels.

(**a**) NO (**b**) N_2O

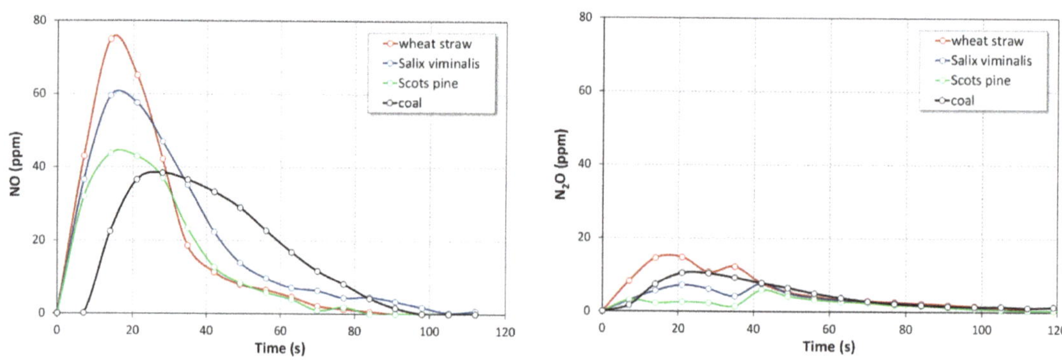

Figure 1. Impact of fuel type on the instantaneous NO (**a**) and N_2O (**b**) concentrations during air combustion.

Nitrous oxide emissions (Figure 1b) were about four times lower than NO emissions, which indicates that fuel-N was more inclined to undergo conversion to NO than to N_2O during combustion in air. The N_2O formed, compared to NO, was also more readily reduced to N_2 by char or CO [27,28]. The highest N_2O concentrations were approximately 15 ppm for wheat straw, while the lowest for Scots pine what was attributed to the nitrogen content in biomass fuels.

Hydrogen cyanide, HCN, and ammonia, NH_3, are major precursors for the formation of NO_x during combustion of solid fuels containing nitrogen. There is a consensus that the conversion of fuel-N to nitrogen oxides proceeds through these intermediates [29]. The instantaneous concentrations of HCN were below 3 ppm and NH_3 was not detected during air combustion of tested fuels.

Figure 2 shows the influence of fuel type on the total emissions of NO and N_2O during combustion in air. In the case of whet straw, NO accounts for approximately 75% of the total NO_x emitted and N_2O for remaining 25%. For *Salix viminalis* and Scots pine, these figures are 85% and 15%, and 72% and 28%, respectively. In the case of bituminous coal, the fraction of NO in the total NO_x is much lower, approximately 40%, and N_2O accounts for remaining 60%. As coal contains less volatile matter than biomass fuels, high emissions of N_2O can be attributed to oxidation and reduction reactions at the surface of char particles.

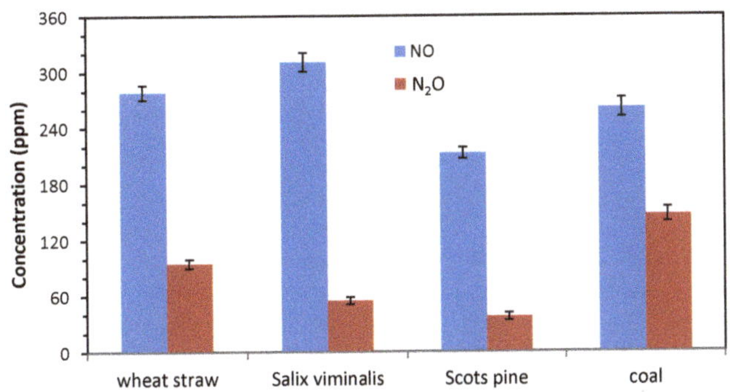

Figure 2. Effect of fuel type on the total NO and N_2O concentrations during air combustion. Vertical bars represent standard deviation.

The split of NO and N_2O emissions between volatile-N and char-N oxidation during air-combustion of wheat straw is shown in Figure 3. Oxidation of volatile matter accounts for approximately 80% of total NO formed and char oxidation contributed only 20%. In the case of N_2O, contributions of volatile-N and char-N are similar and are approximately 51% and 49%.

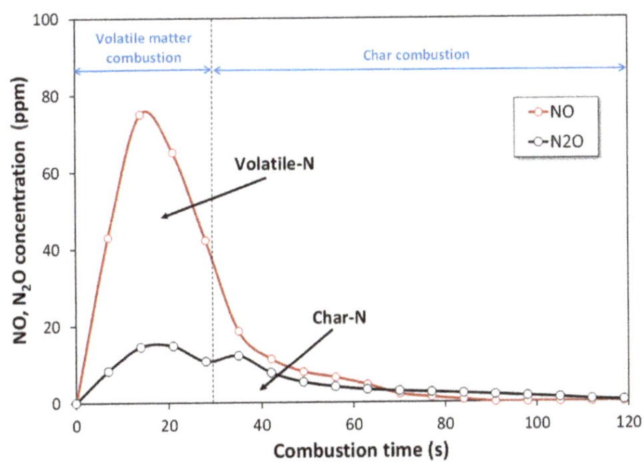

Figure 3. The instantaneous concentrations of NO and N_2O during air-combustion of wheat straw.

Oxidation of volatile-N was responsible for 75% of the total NO formed for *Salix viminalis*, 85% for Scots pine and 65% for coal (Figure 4). Volatile-N contributed 48% to the total N_2O formed during combustion of *Salix viminalis* and 32% during combustion of Scots pine and coal.

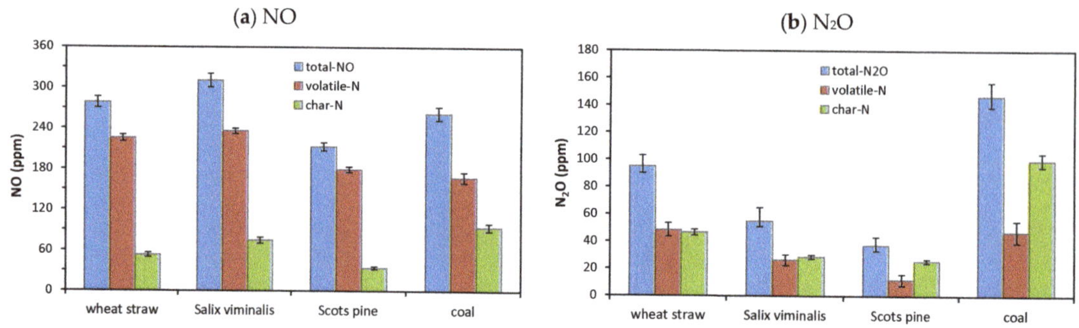

Figure 4. Effect of fuel type on NO (**a**) and N_2O (**b**) concentrations during air combustion. Vertical bars represent standard deviation.

4.1.2. SO_2 Emissions

In the air-combustion process, the sulfur contained in the fuel is oxidized mostly to SO_2 with only a small fraction converted to SO_3 [2]. Biomass fuels contain much less sulfur than coals, therefore, their combustion or co-combustion with coal produces lower emissions of SO_2 [7,8,30]. Figures 5 and 6 show the time-resolved concentrations of SO_2 and the total SO_2 emissions for tested fuels. Emissions of SO_2 from combustion of the tested biomass were much lower than those from the combustion of coal. The highest instantaneous concentrations of SO_2 were observed during combustion of volatiles and they were approximately 58 ppm for coal and less than 9 ppm for biomass fuels.

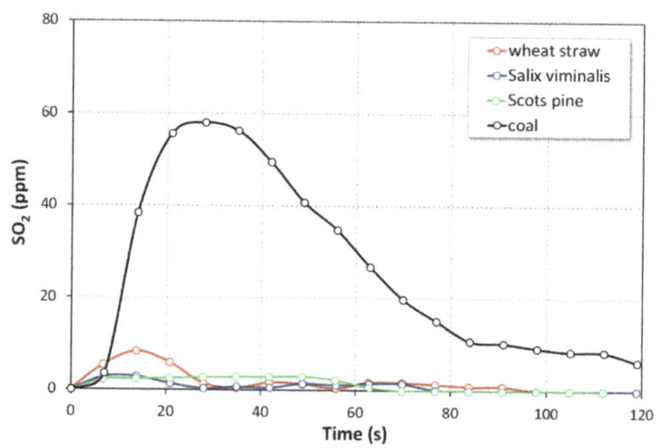

Figure 5. The instantaneous concentrations of SO_2 during air combustion.

Sulphur dioxide is formed through oxidation of sulfur-containing species in the fuel (fuel-S), which are released during devolatilization (volatile-S) and char oxidation (char-S) stages (Figure 6). Oxidation of volatile-S was responsible for 60–65% of the total SO_2 formed for biomass fuels and for only 30% in the case of coal.

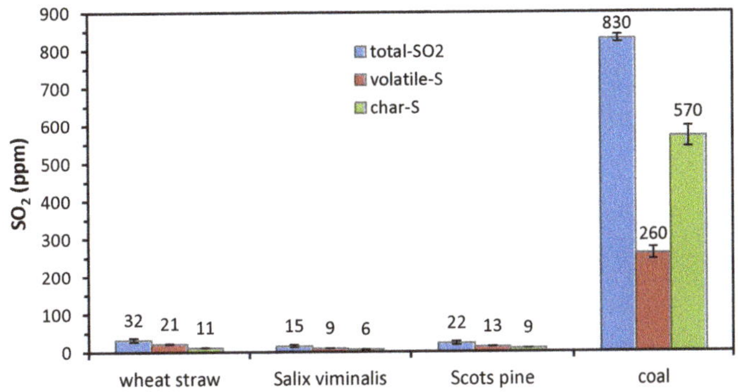

Figure 6. Effect of fuel type on the total SO_2 emissions during air combustion. Vertical bars represent standard deviation.

4.1.3. Emissions of CO

Figure 7 shows carbon monoxide emissions during air combustion of tested fuels. The instantaneous concentrations of CO for the combustion of biomass fuels were much lower than those for the combustion of reference coal. This can be attributed to the lower carbon content (Table 2) of biomass fuels. Release times of CO were 130–140 s for biomass and much longer, approximately 650 s, for coal. The highest CO emissions were detected during the combustion of char, both for biomass and coal. During air-combustion, the highest instantaneous concentration of CO were approximately 22 ppm and 19 ppm for coal and wheat straw, respectively.

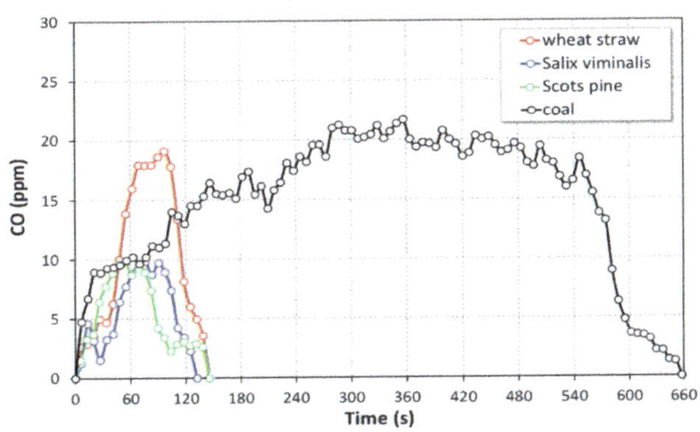

Figure 7. Effect of fuel type on the instantaneous CO concentrations during air combustion.

4.2. Oxy-Fuel Combustion

4.2.1. NO_x Emissions

Figures 8 and 9 show the time-resolved emissions of NO and N_2O during oxy-fuel combustion of all tested fuels. During oxy-combustion, the highest NO emissions were detected during the volatiles combustion after 14 s for Scots pine (Figure 8c), 21 s for wheat straw (Figure 8a) and willow (Figure 8b), and 28 s for coal (Figure 8d). The lowest NO concentrations in a flue gas were observed during combustion in 21% O_2/79% CO_2 environment for all fuels, due to the lower temperature of fuel particle [23,24]. During oxy-21 combustion of wheat straw (Figure 8a), the highest NO concentration was approximately 50 ppm and it was 25 ppm lower than that in air combustion (Figure 1a). Concentrations of NO increased with increasing O_2 content in the O_2/CO_2 mixture. The highest NO concentrations were detected in 40% O_2/60% CO_2 atmosphere and were approximately

95 ppm and 75 ppm for wheat straw and *Salix viminalis*, respectively. The maximum value of NO concentration for wheat straw was about twice as high in oxy-40 than that in oxy-21.

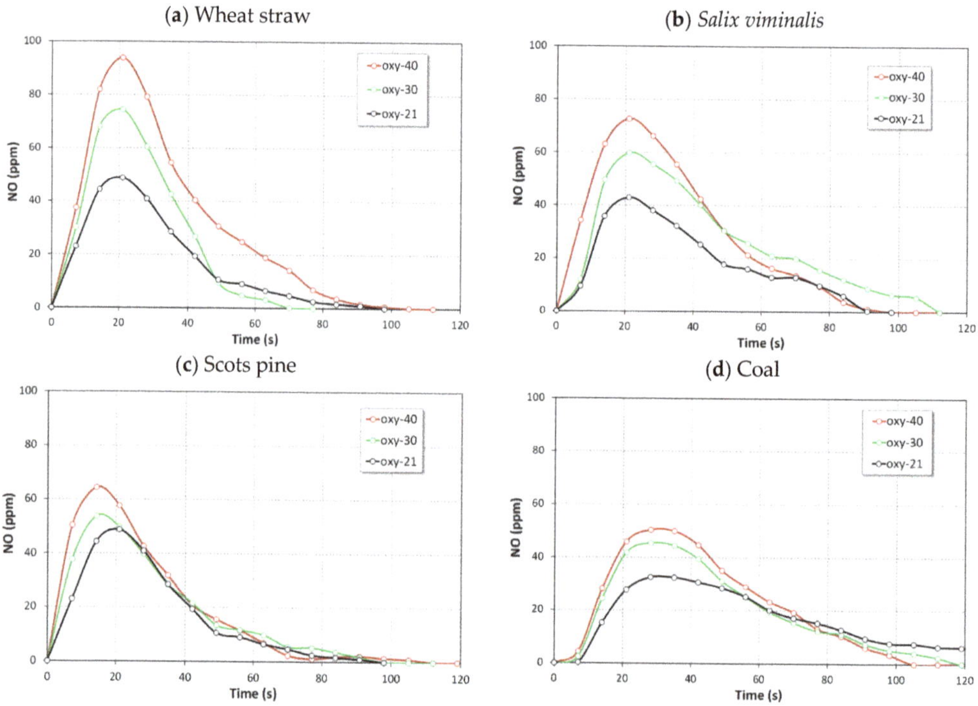

Figure 8. The instantaneous concentrations of NO during oxy-combustion of tested fuels.

Emissions of N_2O were significantly lower in comparison with NO emissions. Nitrous oxide was formed simultaneously with NO, which implies that formation of N_2O proceeded by direct oxidation of fuel nitrogen rather than by the reduction of NO. The highest N_2O concentrations were detected during volatile matter combustion in the oxy-21 atmosphere. The highest value of N_2O concentration, approximately 18 ppm, was observed for the combustion of wheat straw (Figure 9a). As the oxygen concentration in O_2/CO_2 atmosphere increased, N_2O concentrations decreased. The lowest N_2O emissions (below 5 ppm) were observed for oxy-combustion of Scots pine, the fuel with the lowest nitrogen content.

Emissions of NO_2 were only detected during char combustion in oxy-21 atmosphere. The highest instantaneous NO_2 concentration did not exceed 6 ppm for coal and 4 ppm for wheat straw. Figure 10 shows total NO, N_2O and NO_2 emissions during combustion in the oxy-21 atmosphere. Total emissions of NO_2 were approximately 20 ppm and 10 ppm for coal and wheat straw, respectively. This finding is consistent with the results obtained by Lasek et al. [31].

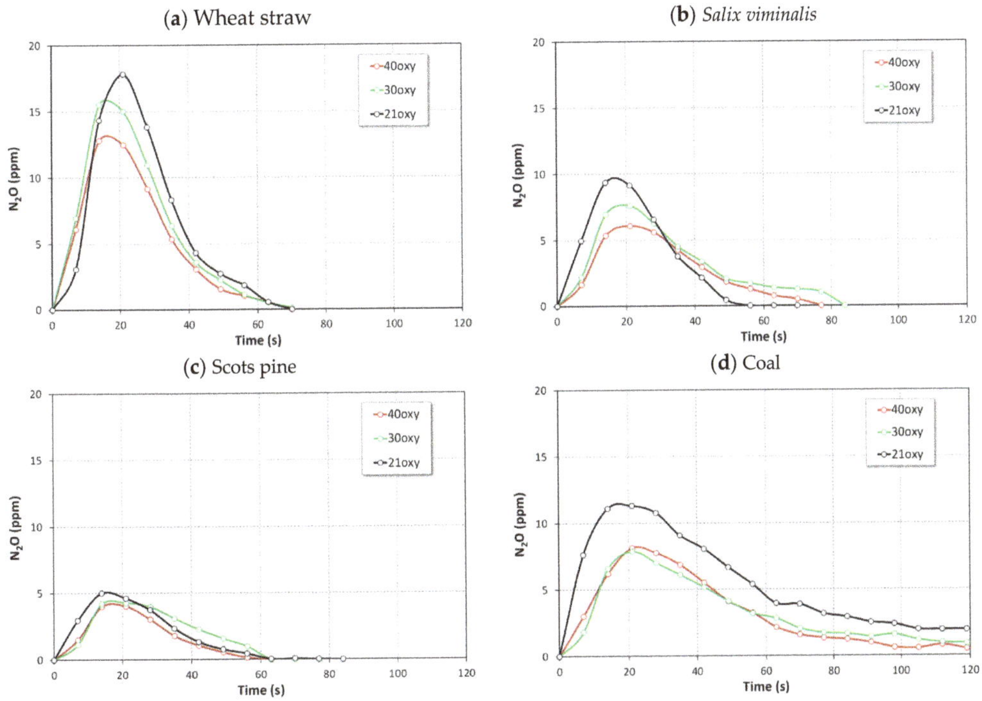

Figure 9. The instantaneous concentrations of N_2O during oxy-combustion of tested fuels.

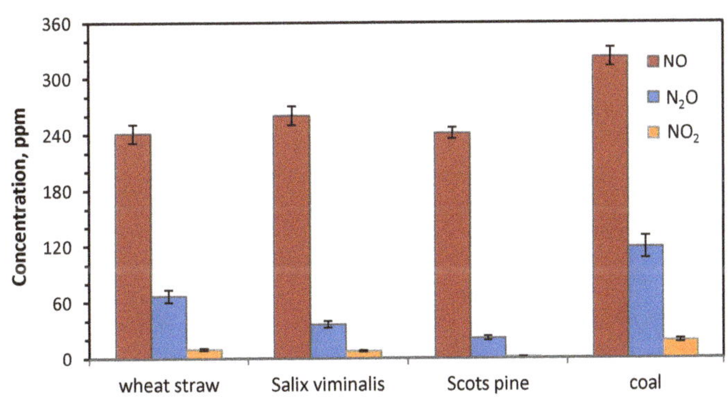

Figure 10. Total emissions of NO, N_2O and NO_2 during combustion in oxy-21 atmosphere. Vertical bars represent standard deviation.

Figure 11 shows the split of NO and N_2O emissions between volatile-N and char-N oxidation during burning of biomass fuels and coal in all atmospheres. Oxidation of volatile matter accounts for approximately 75% of total NO formed and char oxidation contributed 25% for wheat straw in the oxy-21 atmosphere while for coal 40% and 60%. Combustion of wheat straw in the oxy-40 atmospheres caused oxidation of volatile-N for approximately 60% of total NO formed, and char oxidation contributed 40% for wheat straw (68% and 32% for *Salix viminalis*; 80% and 20% for Scots pine; 60% and 40% for coal).

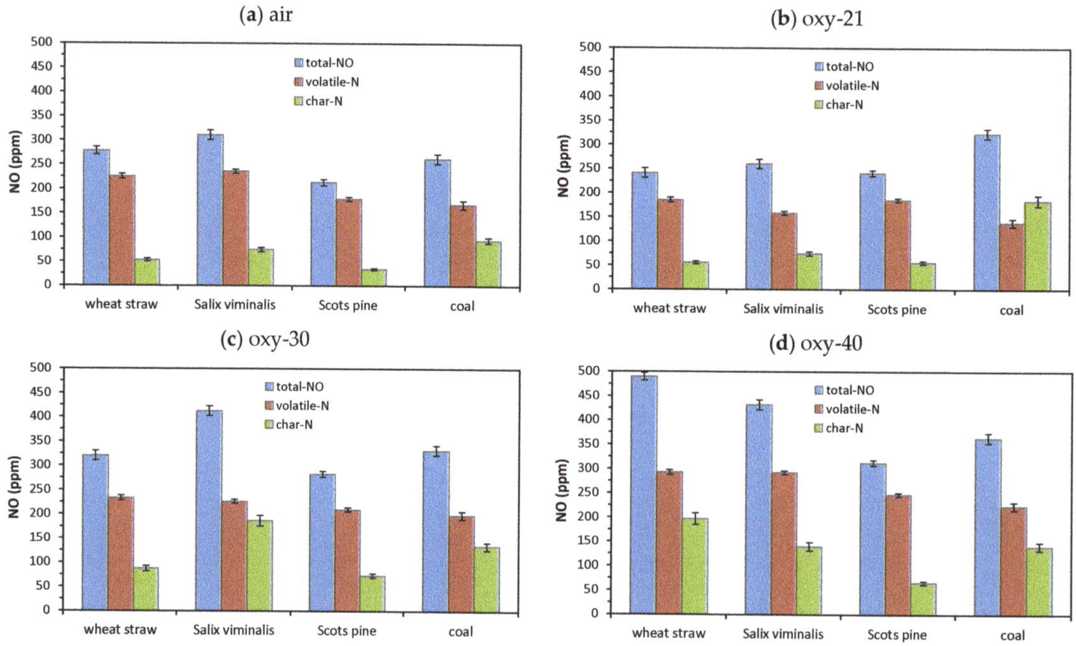

Figure 11. Effect of oxidizing atmosphere on NO emissions during combustion. Vertical bars represent standard deviation.

In the case of N_2O (Figure 12), during oxy-combustion, contributions of volatile-N and char-N are completely different than that in air-combustion. Oxidation of volatile matter accounts for approximately 85% of total N_2O formed, and char oxidation contributed 15% for wheat straw and Scots pine in the oxy-21 atmosphere. (93% and 7% for *Salix viminalis*; 80% and 20% for Scots pine; 51% and 49% for coal).

Combustion of wheat straw in the oxy-40 atmospheres caused oxidation of volatile matter for approximately 77% of total N_2O formed and, char oxidation contributed 23% for wheat straw (75% and 25% for Salix viminalis; 90% and 10% for Scots pine; 70% and 30% for coal).

The influence of oxygen concentration in the O_2/CO_2 mixture on the total emissions of NO is shown in Figure 13. As expected, emissions of NO increase with increasing O_2 content in the mixture, particularly for biomass fuels. The emission of NO also increased with the particle temperature due to the decrease in CO emission. Higher particle temperatures helped the oxidation of NCO to NO, too [29]. Moreover, higher concentration of O_2 in the riser enhance the combustion of volatile matter and char and lead to an increase in NO formation. This remark has been confirmed by the results obtained by Czakiert et al. [32] and Jankowska et al. [33].

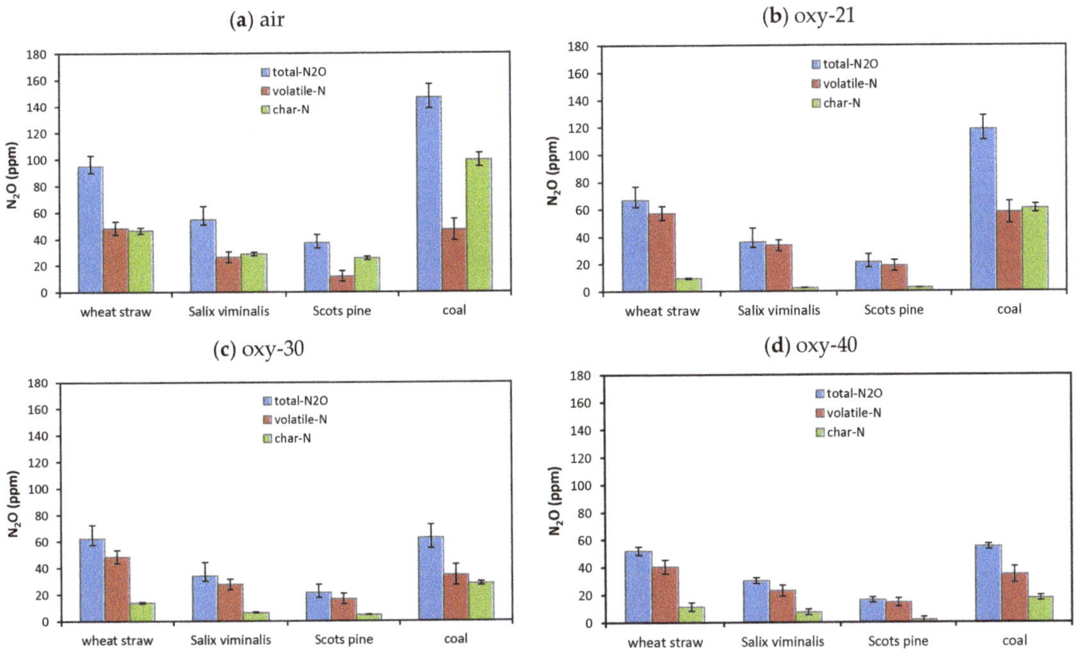

Figure 12. Effect of oxidizing atmosphere on N_2O emissions during combustion. Vertical bars represent standard deviation.

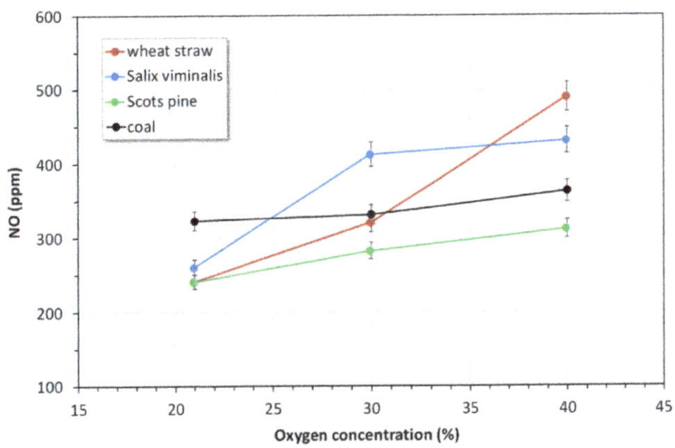

Figure 13. Influence of oxygen concentration on the total emission of NO during oxy-combustion. Vertical bars represent standard deviation.

The opposite trend can be observed for N_2O emissions (Figure 14). As the oxygen concentration in O_2/CO_2 mixture increases, N_2O concentrations in a flue gas decrease, mainly for coal.

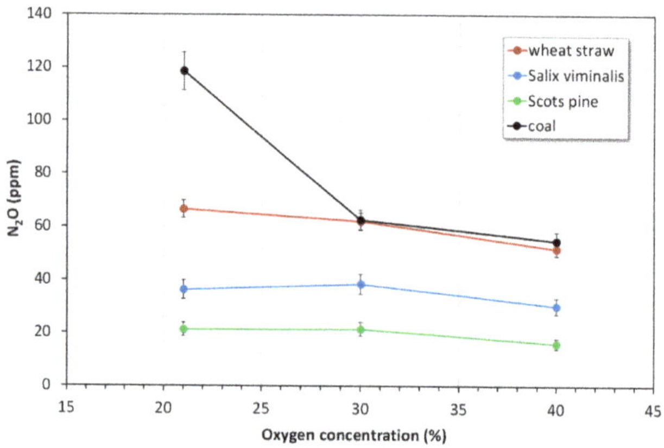

Figure 14. Effect of the oxygen concentration on the total emissions of N_2O during oxy-combustion. Vertical bars represent standard deviation.

Hydrogen cyanide is a main precursor of N_2O formation. It should be mentioned that HCN is a very toxic gas (approximately 20 times more toxic than CO) that already affects human beings at ppm levels [31]. Higher concentrations of HCN were generated during oxy-fuel combustion. Emissions of HCN were the highest in oxy-21 atmosphere, particularly during char combustion. The highest instantaneous concentrations of HCN did not exceed 6 ppm both for biomass fuels and coal. Similar results were reported in our previous study [22] and by Lasek et al. [31].

The highest NH_3 emissions were observed during combustion in oxy-30 and oxy-21 atmospheres. However, the instantaneous concentrations of NH_3 were very low; they did not exceed 3 ppm for all fuels. High concentrations of CO_2 inhibited oxidation of HCN and NH_3 to NO_x.

4.2.2. Emissions of SO_2

The SO_2 emissions during biomass fuels and coal combustion in various environments are shown in Figure 15. The instantaneous SO_2 concentrations during oxy-combustion of biomass were much lower than those for the combustion of reference coal. The times of SO_2 release were 40–90 s for biomass and 380–580 s for coal in all O_2/CO_2 mixtures. The highest emissions of SO_2 were observed during the combustion of volatiles, for both biomass and coal. Compared with air-combustion, burning both biomass fuels and coal in oxy-21 atmosphere caused a minor increase in emissions of SO_2. The highest SO_2 emission in oxy-21 mixture was approximately 62 ppm for coal. An increase in the oxygen inlet concentration caused higher emissions of SO_2. The highest SO_2 emissions were detected in oxy-40 atmosphere and was 86 ppm for coal. The highest concentrations of SO_2 in oxy-40 for biomass fuels were 21 ppm, 13 ppm and 10 ppm for wheat straw, *Salix viminalis* and Scots pine, respectively.

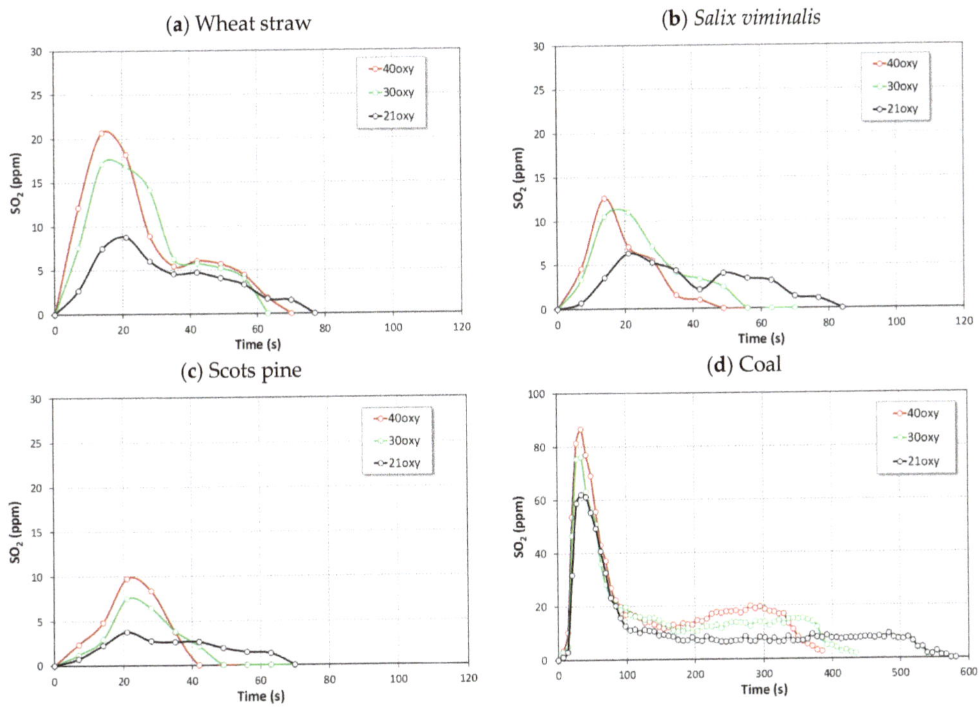

Figure 15. The instantaneous concentrations of SO_2 during oxy-combustion of the tested fuels.

Figure 16 shows the influence of the oxygen concentration on the total emission of SO_2. It can be seen that an increase in the oxygen inlet concentration resulted in higher SO_2 emissions, particularly for coal. Combustion of wheat straw in the oxy-40 atmosphere caused twice as higher emission of SO_2 than that in oxy-21. A similar trend for coal and biomass fuels during oxy-combustion was found in references [19,31,34,35].

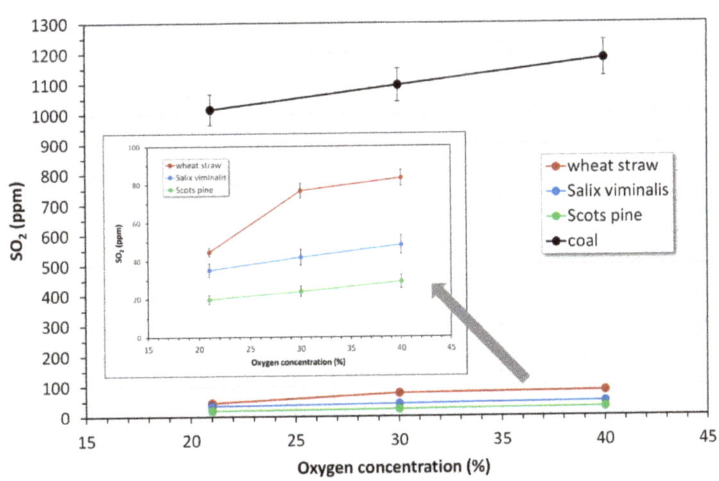

Figure 16. Influence of the oxygen concentration on total emissions of SO_2 during oxy-combustion. Vertical bars represent standard deviation.

4.2.3. CO Emissions

Concentrations of CO during oxy-fuel combustion are shown in Figures 17 and 18. These results reveal a significant impact of the composition of oxidizing gas mixture on CO emissions. Combustion of biomass or coal in the oxy-21 mixture caused a strong increase in CO concentrations in comparison with air-combustion (see Figure 3). The highest CO emissions were approximately 950 ppm and 760 ppm for wheat straw and coal, respectively. This phenomenon can be explained by the influence of three chemical reactions [2,3,31,32]: Boudouard's reaction Equation (1), the reaction of CO_2 with H radical Equation (2) and the dissociation of CO_2 Equation (3):

$$CO_2 + C \leftrightarrow 2CO \quad (1)$$

$$CO_2 + H \leftrightarrow CO + OH \quad (2)$$

$$CO_2 \leftrightarrow CO + \frac{1}{2}O_2 \quad (3)$$

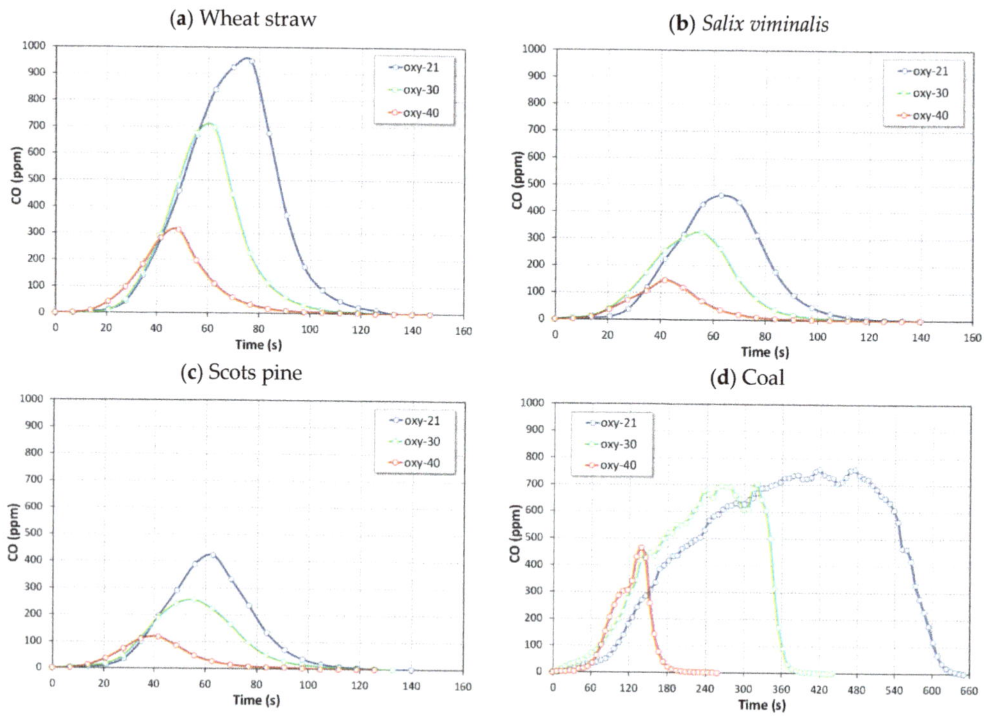

Figure 17. Effect of oxygen concentration on the instantaneous CO concentrations during oxy-combustion.

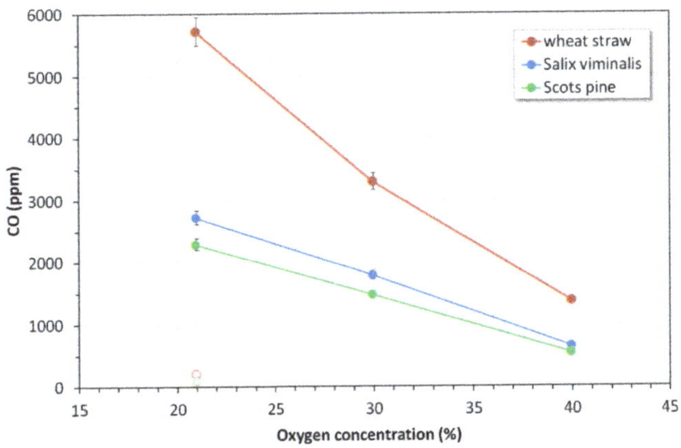

Figure 18. Influence of the oxygen concentration on total emissions of CO during combustion of biomass fuels. Open points represent combustion in air. Vertical bars represent standard deviation.

Reaction (1) seems to play a leading role in the conversion of CO under conditions typical for fluidized-bed combustion [31]. As oxygen concentration increased, the emission of CO was reduced. In oxy-40 environment, the highest CO emissions were three times lower for biomass fuels than that in the oxy-21 mixture. The highest instantaneous CO concentrations were 760 ppm, 700 ppm and 465 ppm in oxy-21, oxy-30 and oxy-40, respectively. The analogous trends were reported in references [31,36–38].

Moreover, the reduction of NO to N_2 in the presence of an active char and higher concentrations of CO is enhanced through the heterogeneous NO reduction reaction [37,39]:

$$CO + NO \leftrightarrow CO_2 + \frac{1}{2}N_2 \quad (4)$$

According to results presented by Lupianez et al. [14–16], a high level of CO in the flue gas is a result of incomplete combustion, which is a symptom of lower combustion efficiency. Thus, in this case, the reduction of NO to N_2 is a consequence of such deficiency (Figure 19).

The results of this study show that the tested biomass fuels are ideal renewable energy resources both in conventional and oxy-fuel conditions with a minor potential for environmental pollution. According to the EU's Large Combustion Plant Directive, emission limits for NO (as NO_2), SO_2 and CO are 200 mg/$m^3{}_n$ (dry gas at 6% O_2), and 20 mg/$m^3{}_n$ for particulate matter (fly ash). In the case of coal-fired CFB boilers, these limits can only be met with De-NO_x (selective non-catalytic reduction) and De-SO_x (in situ SO_2 capture) systems. Biomass fuels usually contain much less nitrogen, sulfur and mineral matter than coals. Since the magnitude of NO_x and SO_2 emissions is strongly related to the content of N and S in the fuel, a proper combination of combustion technique, operating conditions, and biomass fuel can deliver a combustion system with significantly reduced requirements for De-NO_x and De- SO_x systems or even eliminate them.

These experimental results can be used to optimize the combustion process of biomass fuels in the CFB oxy-combustion technology. The data can also be useful in mathematical modeling of biomass combustion both in conventional and oxy-combustion processes.

(**a**) Oxy-21

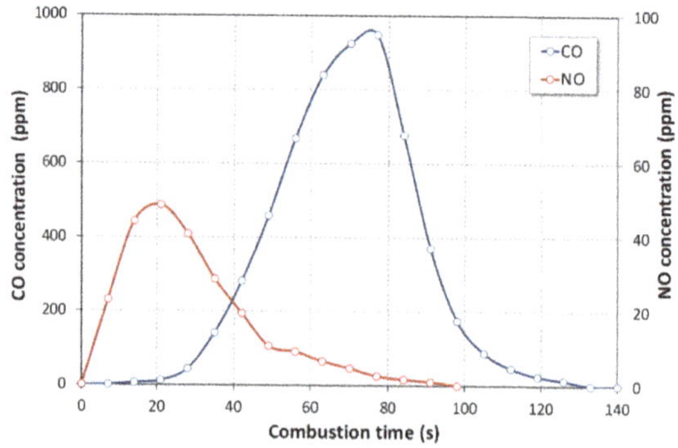

(**b**) Oxy-30

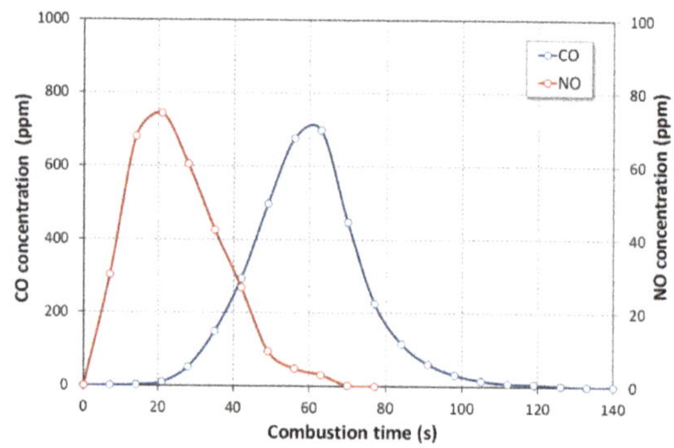

(**c**) Oxy-40

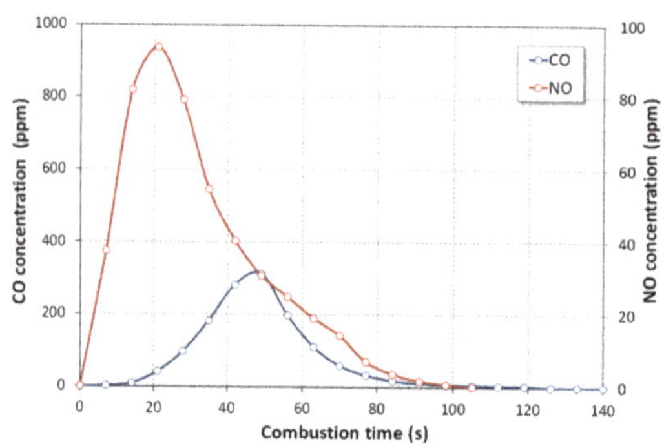

Figure 19. NO and CO (ppm) profiles during combustion of wheat straw under oxy-combustion.

5. Conclusions

Three kinds of biomass and bituminous coal were burned in a laboratory-scale CFB combustor, and the influence of fuel type and oxidizing atmosphere on pollutant emissions were studied. Pellets were burned in air and oxy-combustion environments (oxy-21, oxy-30 and oxy-40) at 850 °C. The major conclusions from this study are as follows:

1. The instantaneous emissions of SO_2, N_2O and CO for the combustion of biomass fuels in all atmospheres were lower than those for the combustion of bituminous coal.

2. The instantaneous NO concentrations for the combustion of biomass fuels in all atmospheres were higher than those for the combustion of reference coal.
3. Oxidation of nitrogen species released with volatile matter was responsible for high emissions of NO_x during combustion of biomass fuels in air and mixtures of O_2 and CO_2.
4. The highest instantaneous emissions of all gaseous pollutant during combustion of biomass fuels were observed for agro biomass (wheat straw) in both air and oxy-fuel atmospheres.
5. The lowest emissions of all gaseous pollutant during combustion of biomass fuels were observed for woody biomass (Scots pine) in both air and oxy-fuel atmospheres.
6. The lowest NO emissions and the highest N_2O emissions for tested fuels were detected in oxy-21 atmosphere owing to lower temperature of fuel particles.
7. The instantaneous CO emissions for the combustion of biomass in all atmospheres were much lower than those for the combustion of reference coal.
8. Oxy-combustion of biomass or coal caused a significant increase in emissions of CO in comparison with air-combustion. This observation can be attributed to $CO_2 + C_{(char)}$ reaction (the Boudouard reaction).
9. Combustion of biomass in oxy-30 or oxy-40 atmospheres causes an increase in SO_2 and NO emissions and a reduction in of CO and N_2O emissions.
10. Considering all pollutant emissions during oxy-combustion of biomass fuels, the optimal atmosphere is oxy-30.

Author Contributions: Conceptualization, M.K.-G., A.K.-K. and A.L.; methodology, M.K.-G.; formal analysis, M.K.-G. and A.K.-K.; investigation, M.K.-G. and A.K.-K.; writing—original draft preparation, M.K.-G.; writing—review and editing, M.K.-G. and A.L.; project administration, M.K.-G.; funding acquisition, M.K.-G. All authors have read and agreed to the published version of the manuscript.

Funding: Funds to cover costs of publishing in an open access journal were provided by the statute subvention of Czestochowa University of Technology, Faculty of Mechanical Engineering and Computer Science, Department of Thermal Machinery.

Institutional Review Board Statement: Not applicable.

Informed Consent Statement: Not applicable.

Data Availability Statement: Not applicable.

Acknowledgments: The results presented in this paper were obtained from research work co-financed by the National Science Centre (Poland) based on the decision No. DEC-2012/07/B/ST8/03730. The support is gratefully acknowledged.

Conflicts of Interest: The authors declare no conflict of interest.

References

1. Yadav, S.; Mondal, S.S. A review on the process and prospects of oxy-fuel carbon capture and sequestration (CCS) technology. *Fuel* **2022**, *308*, 122057. [CrossRef]
2. Stanger, R.; Wall, T.; Spörl, R.; Paneru, M.; Grathwohl, S.; Weidmann, M.; Scheffknecht, G.; McDonald, D.; Myöhänen, K.; Ritvanen, J.; et al. Oxyfuel combustion for CO_2 capture in power plant. *Int. J. Greenh. Gas Control* **2015**, *40*, 55–125. [CrossRef]
3. Toftegaard, M.B.; Brix, J.; Jensen, P.A.; Glarborg, P.; Jensen, A.D. Oxy-fuel combustion of solid fuels. *Prog. Energy Combust. Sci.* **2010**, *36*, 581–625. [CrossRef]
4. Liu, Q.; Shi, Y.; Zhong, W.; Yu, A. Co-firing coal and biomass in oxy-fuel fluidized bed for CO_2 capture: A review of recent advances. *Chin. J. Chem. Eng.* **2019**, *27*, 2261–2272. [CrossRef]
5. Duan, L.; Duan, Y.; Zhao, C.; Anthony, E.J. NO emission during co-firing coal and biomass in an oxy-fuel circulating fluidized bed combustor. *Fuel* **2015**, *150*, 8–13. [CrossRef]
6. Zhang, X.; Meloni, S. *Technology Developments in the Cofiring of Biomass*; IEA, Clean Coal Centre: London, UK, 2020.
7. Williams, A.; Jones, J.M.; Ma, L.; Pourkashanian, M.M. Pollutants from the combustion of solid biomass fuels. *Prog. Energy Combust. Sci.* **2012**, *38*, 113–137. [CrossRef]
8. Varol, M.; Symonds, R.; Anthony, E.J.; Lu, D.; Jia, L.; Tan, Y. Emissions from co-firing lignite and biomass in an oxy-fired CFBC. *Fuel Processing Technol.* **2018**, *173*, 126–133. [CrossRef]

9. Singh, R.I.; Kumar, R. Current status and experimental investigation of oxy-fired fluidized bed. *Renew. Sustain. Energy Rev.* **2016**, *61*, 398–420. [CrossRef]
10. Lupion, M.; Alvarez, I.; Otero, P.; Kuivalainen, R.; Lantto, J.; Hotta, A.; Hack, H. 30 MWth CIUDEN oxy-CFB boiler—First experience. *Energy Procedia* **2013**, *37*, 6179–6188. [CrossRef]
11. Anthony, E.J.; Hack, H. Oxy-fired fluidized bed combustion: Technology prospects and new developments. In *Fluidized Bed Technologies for Near-Zero Emission Combustion and Gasification*; Scala, F., Ed.; Woodhead Publishing: London, UK, 2013; pp. 867–894.
12. Gao, Y.; Tahmasebi, A.; Dou, J.; Yu, J. Combustion characteristics and air pollutant formation during oxy-fuel co-combustion of microalgae and lignite. *Bioresour. Technol.* **2016**, *207*, 276–284. [CrossRef] [PubMed]
13. Tan, T.; Jia, L.; Wu, Y. Some combustion characteristics of biomass and coal cofiring under oxy-fuel conditions in a pilot-scale circulating fluidized combustor. *Energy Fuels* **2013**, *27*, 7000–7007. [CrossRef]
14. Lupiáñez, C.; Carmen Mayoral, M.; Guedea, I.; Espatolero, S.; Díez, L.I.; Laguarta, S.; Manuel Andrés, J. Effect of co-firing on emissions and deposition during fluidized bed oxy-combustion. *Fuel* **2016**, *184*, 261–268. [CrossRef]
15. Lupiáñez, C.; Carmen Mayoral, M.; Díez, L.I.; Pueyo, E.; Espatolero, S.; Díez, L.I.; Manuel Andrés, J. The role of limestone during fluidized bed oxy-combustion of coal and biomass. *Appl. Energy* **2016**, *184*, 670–680. [CrossRef]
16. Lupiáñez, C.; Carmen Mayoral, M.; Díez, L.I.; Pueyo, E.; Espatolero, S.; Manuel Andrés, J. On the oxy-combustion of lignite and corn stover in a lab-scale fluidized bed reactor. *Biomass Bioenergy* **2017**, *96*, 152–161. [CrossRef]
17. Pu, G.; Zang, H.; Du, J.; Zhang, X. Study on NO emission in the oxy-fuel combustion of cofiring coal and biomass in a bubbling fluidized bed combustor. *BioResources* **2017**, *12*, 1890–1902. [CrossRef]
18. Wang, X.; Ren, Q.; Li, W.; Li, H.; Li, S.; Lu, Q. Nitrogenous gas emissions from coal/biomass co-combustion under a high oxygen concentration in a circulating fluidized bed. *Energy Fuels* **2017**, *31*, 3234–3242. [CrossRef]
19. Sher, F.; Pans, M.A.; Sun, C.; Snape, C.; Liu, H. Oxy-fuel combustion study of biomass fuels in a 20 kWth fluidized bed combustor. *Fuel* **2018**, *215*, 778–786. [CrossRef]
20. Nguyen, H.K.; Moon, J.-H.; Jo, S.-H.; Park, S.J.; Seo, M.W.; Ra, H.W.; Yoon, S.-J.; Yoon, S.-M.; Song, B.; Lee, U.; et al. Oxy-combustion characteristics as a function of oxygen concentration and biomass co-firing ratio in a 0.1 MWth circulating fluidized bed combustion test-rig. *Energy* **2020**, *196*, 117020. [CrossRef]
21. Kosowska-Golachowska, M.; Luckos, A.; Kijo-Kleczkowska, A.; Musiał, T.; Wolski, K.; Środa, K. Gaseous emissions during oxy-fuel combustion of sewage sludge in a circulating fluidized bed. *Powder Technol.* **2020**, *371*, 209–216. [CrossRef]
22. Kosowska-Golachowska, M.; Luckos, A.; Musiał, T.; Kijo-Kleczkowska, A.; Wolski, K.; Środa, K. NO_x and N_2O emissions during oxy-fuel combustion of bituminous coal and lignite in a circulating fluidized bed combustor. In *Proceedings of the 13th International Conference on Fluidized Bed Technology, Vancouver, BC, Canada, 10–14 May 2020*; Bi, X., Briens, C., Ellis, N., Wormsbecker, M., Eds.; GLAB Reactor and Fluidization Technologies: Vancouver, BC, Canada, 2021; pp. 587–592.
23. Kosowska-Golachowska, M.; Kijo-Kleczkowska, A.; Luckos, A.; Wolski, K.; Musiał, T. Oxy-combustion of biomass in a circulating fluidized bed. *Arch. Thermodyn.* **2016**, *37*, 17–30. [CrossRef]
24. Kosowska-Golachowska, M.; Otwinowski, H.; Wolski, K.; Kijo-Kleczkowska, A.; Musiał, T.; Środa, K.; Richter, D. Oxy-fuel combustion of wheat straw pellets in a lab-scale fluidized bed combustor. In *Renewable Energy Sources: Engineering, Technology, Innovation*; Springer Proceedings in Energy; Mudryk, K., Werle, S., Eds.; Springer: Cham, Switzerland, 2018; pp. 375–386.
25. Kijo-Kleczkowska, A.; Środa, K.; Kosowska-Golachowska, M.; Musiał, T.; Wolski, K. Combustion of pelleted sewage sludge with reference to coal and biomass. *Fuel* **2016**, *170*, 141–160. [CrossRef]
26. Shah, I.A.; Gou, X.; Zhang, Q.; Wu, J.; Wang, E.; Liu, Y. Experimental study on NO_x emission characteristics of oxy-biomass combustion. *J. Clean. Prod.* **2018**, *199*, 400–410. [CrossRef]
27. Chyang, C.S.; Wu, K.T.; Lin, C.S. Emission of nitrogen oxides in a vortexing fluidized bed combustor. *Fuel* **2007**, *86*, 234–243. [CrossRef]
28. Liu, X.; Luo, Z.; Yu, C. Conversion of char-N into NO_x and N_2O during combustion of biomass char. *Fuel* **2019**, *242*, 389–398. [CrossRef]
29. Liu, X.; Yang, X.; Xie, G.; Yu, Y. NO emission characteristic during fluidized combustion of biomass with limestone addition. *Fuel* **2021**, *291*, 120264.
30. Roknia, E.; Renb, X.; Panahia, A.; Levendis, Y.A. Emissions of SO_2, NO_x, CO_2, and HCl from co-firing of coals with raw and torrefied biomass fuels. *Fuel* **2018**, *211*, 363–374. [CrossRef]
31. Lasek, J.A.; Janusz, M.; Zuwała, J.; Głód, K.; Iluk, A. Oxy-fuel combustion of selected solid fuels under atmospheric and elevated pressures. *Energy* **2013**, *62*, 105–112. [CrossRef]
32. Czakiert, T.; Sztekler, K.; Karski, S.; Markiewicz, D.; Nowak, W. Oxy-fuel circulating fluidized bed combustion in a small pilot-scale test rig. *Fuel Processing Technol.* **2010**, *91*, 1617–1623. [CrossRef]
33. Jankowska, S.; Czakiert, T.; Krawczyk, G.; Borecki, P. model of fuel combustion in oxygen-enriched CFB environment. *Fuel Processing Technol.* **2010**, *91*, 290–295.
34. Jia, L.; Tan, Y.; Anthony, E.J. Oxy-emissions of SO_2 and NO_x during fuel CFB combustion tests in a mini-circulating fluidized bed combustion reactor. *Energy Fuel* **2010**, *24*, 910–915. [CrossRef]
35. Ahn, J.; Okerlund, R.; Fry, A.; Eddings, E.G. Sulfur trioxide formation during oxy-coal combustion. *Int. J. Greenh. Gas Control* **2011**, *5*, S127–S135. [CrossRef]

36. Krzywanski, J.; Czakiert, T.; Muskala, W.; Sekret, R.; Nowak, W. Modeling of solid fuels combustion in oxygen-enriched atmosphere in circulating fluidized bed boiler: Part 1. The mathematical model of fuel combustion in oxygen-enriched CFB environment. *Fuel Processing Technol.* **2010**, *91*, 290–295. [CrossRef]
37. Normann, F.; Andersson, K.; Leckner, B.; Johnsson, F. Emission control of nitrogen oxides in the oxy-fuel process. *Prog. Energy Combust. Sci.* **2009**, *35*, 385–397. [CrossRef]
38. Meng, X.; Rokni, E.; Zhou, W.; Qi, H.; Sun, R.; Levendis, Y.A. Emissions from oxy-combustion of raw and torrefied biomass. *J. Energy Resour. Technol.* **2020**, *142*, 122307. [CrossRef]
39. Allen, D.; Hayhurst, A.N. The chemical reactions of nitric oxide with solid carbon and catalytically with gaseous carbon monoxide. *Fuel* **2015**, *142*, 260–267. [CrossRef]

energies MDPI

Article

Swirling Flame Combustion of Heavy Fuel Oil Blended with Diesel: Effect of Asphaltene Concentration

Xinyan Pei [1,*], Hongyu Tian [1] and William L. Roberts [2]

1 Institute for Aero Engine, Tsinghua University, Beijing 100084, China
2 Clean Combustion Research Center, King Abdullah University of Science and Technology, Thuwal 23955-6900, Saudi Arabia
* Correspondence: peixinyan@mail.tsinghua.edu.cn

Abstract: Heavy fuel oil has an energy density content comparable to distillate fuels but a very high viscosity that necessitates extra heating before spray combustion inside a boiler. Heavy fuel oil is also characterized by high asphaltenes, carbon residues, trace metals, such as vanadium and nickel, fuel-bound nitrogen, and sulfur. Asphaltenes are heavy polycyclic aromatic compounds with embedded heteroatoms and significantly affect the physico-chemical properties of heavy fuels; this makes them very difficult to burn and leads to the formation of large cenospheres (lightweight, inert, hollow spheres), due to an inefficient burning process. The primary goal of this study is to experimentally investigate the influence of the asphaltene concentration on the combustion of HFO under a swirling flame, finally reducing gaseous and solid pollution. We tested HFO samples containing asphaltene concentrations of 4, 6, 8, 16, and 24 wt.%, prepared by blending the light oil of diesel and pure asphaltenes with HFO. This work provides quantitative information about the effect of different asphaltene contents on the fuel properties of viscosity, density, heating value, thermogravimetry with air and N_2, and swirling flame combustion characteristics, including the analysis of gaseous and solid particle emissions. The results indicated that the high asphaltene content in the fuel was the critical factor for the high viscosity and incomplete combustion and also increased the density of the fuel sample. Reducing the asphaltene content in HFO improves its spray characteristics and combustion performance and reduces the solid emissions containing sulfur and metal elements.

Keywords: heavy fuel oil; asphaltene; swirling flame; blend fuel; fuel characteristics; pollutant emissions

Citation: Pei, X.; Tian, H.; Roberts, W.L. Swirling Flame Combustion of Heavy Fuel Oil Blended with Diesel: Effect of Asphaltene Concentration. *Energies* **2022**, *15*, 6156. https://doi.org/10.3390/en15176156

Academic Editors: Monika Kosowska-Golachowska and Tomasz Czakiert

Received: 27 July 2022
Accepted: 19 August 2022
Published: 24 August 2022

1. Introduction

As to the wide application of heavy fuel oil (HFO) in utility boilers, industrial furnaces, and marine engines, the burning of HFO has been broadly tested in laboratory research and engineering development. With the gradual reduction of conventional light and medium oil resources, efficiently and economically recovering vast reserves of unconventional heavy oil and asphalt has attracted increasing attention. HFO has an energy density content similar to distilled fuel, but it has a very high viscosity and requires additional heating before spray combustion in the boiler [1]. The specific characteristics of HFO are the high contents of asphaltenes, carbon residues, trace metals, such as vanadium and nickel, and fuel-bound nitrogen and sulfur. Asphaltenes are heavy polycyclic aromatic compounds with embedded heteroatoms, making the fuel difficult to burn, and the inefficient combustion process leads to the formation of large cenospheres (lightweight, inert, hollow spheres).

Asphaltenes are compounds in crude oil, heavy fuels (high-boiling and nonboiling petroleum fractions), and oil sand bitumen. They are insoluble in n-alkanes (e.g., n-heptane) and only soluble in aromatic solvents (e.g., toluene). Their chemical structures are still not fully understood, so asphaltenes are characterized by their solubility class instead of their chemical properties. Asphaltenes have a significant impact on the physicochemical properties of heavy fuels and residues. It has been shown that asphaltenes are the most aromatic part of heavy fuels and, therefore, increase the viscosity of the fuel [2]. Thus, HFO with high asphaltene content requires the adding of solvent or the preheating of the fuel to improve fluidity. Asphaltenes in crude oil precipitate and deposit along the walls of the oil reservoir, thereby clogging the oil well and causing flowline fouling problems. For all these reasons, asphaltenes have been called 'bad guys' in petroleum fuels [3].

The well-referred results established by HFO combustion were to distinguish two combustion stages. The first is the liquid phase of evaporating volatile substances, and, then, the solid phase of oxidized coke, carbonaceous cenosphere particles formed during the liquid phase. This liquid combustion stage is a complex process that involves heat and mass transfer and chemical reactions, such as pyrolysis (thermal cracking) and polymerization. The solid combustion phase occurs immediately after the liquid combustion phase and forms a hollow shell cenosphere [4]. Most of the research on HFO combustion reported in the literature was carried out using suspended [5–7] and falling droplet [8–10] techniques, and many researchers used numerical simulations [11–13]. Elbaz et al. [7] studied the formation and oxidation of heavy oil, HFO, and particles produced by the combustion of droplets. They also used a scanning electron microscope (SEM) and an energy dispersive X-ray (EDX) to analyze the microstructure of the particles. The study concluded that the droplet ignition temperature is independent of the droplet size, but the liquid phase ignition delay time and droplet life are proportional to the initial droplet diameter. Kwack et al. [14] burned No. 6 fuel in a closed burner, observed the combustion residue through a scanning electron microscope, and studied the qualitative relationship between the shape of these particles and the temperatures to which they were subjected. Xu et al. [4] divided the process of heavy oil droplet combustion into four steps: ignition delay, flame lifetime, coke glowing delay, and coke ember time. Using the mixed oil of mixed heavy oil residue (HOR) and diesel light oil (LO), they further analyzed their composition characteristics in terms of oil composition and combustion conditions. They found that increasing the temperature of the combustion chamber significantly reduced the ignition delay and the coke luminescence delay, but hardly changed the flame lifetime. Ambalae et al. [15] used a thermogravimetric analyzer (TGA) to obtain information on the pyrolysis and combustion behavior of crude oil (Neilburg) and its asphaltenes. The study found that asphaltenes contributed the most to coke formation among all saturated, aromatic, resin, and asphaltene fractions. They analyzed the temperature rise of whole oil and asphaltenes and conducted isothermal pyrolysis experiments to determine the temperature at which coke formation was maximized. In addition, they obtained isothermal combustion curves of coke derived from whole oil and asphaltenes. Atiku et al. [16] explored the mechanism of forming fine particulate soot and cenospheres and studied the chemical structure of petroleum asphaltenes through pyrolysis technology. Jameel et al. [17] used non-isothermal thermogravimetric analysis (TGA) and Fourier transform infrared (FTIR) spectrometers to study the pyrolysis and combustion of heavy oil in nitrogen and air, respectively, and deeply understood the three stages of heavy oil combustion.

Some work has been conducted on the influences of oil compositions on HFO burning. The effect of oil composition on HFO burning was studied by changing asphaltene content [5,18,19], sulfur [6,20], metal (vanadium) [21], viscosity [22,23], and fuel nitrogen [24]. The viscosity of HFO largely depends on the volume fraction, chemical structure, and physicochemical properties of its asphaltenes. Asphaltenes are the most polar and heaviest components in HFO [25]. Fakher et al. [26] explained the main components of crude oil and its relationship with asphaltenes and the methods for quantifying asphaltenes in crude oils, showing more discussed models for asphaltene modeling, and mentioned the chemistry used to characterize and study asphaltene analysis methods. In addition, they also introduced the methods by which asphaltenes destroy oil recovery. The structure of asphaltenes was studied through atomic force microscopy with atomic resolution imaging and scanning tunnelling microscopy with molecular orbital imaging to study more than 100 asphaltene molecules [27]. Peng et al. [28] conducted experimental and theoretical studies on the specific effects of asphaltene content on the viscosity of heavy oil at different temperatures. They determined four important parameters to characterize the reconstituted heavy oil samples: solvation constant, shape factor, intrinsic viscosity, and maximum filling volume fraction. The study showed, through the results of nonlinear regression, that the state of asphaltene particles in heavy oil changed with the change of asphaltene content and temperature, and this change greatly influenced the viscosity of heavy oil. Bartle [29] showed that asphaltene reduced the ignition delay time of heavy oil but did not affect the burning time of fuel droplets. The reduction in ignition delay was attributed to the volatiles produced by the pyrolysis of asphaltenes.

The current work is dedicated to showing how blending asphaltene influences the combustion of heavy fuel oil. The emission characteristics (gaseous emissions and particulate matter) of different asphaltene constituent oils were tested in the swirling flame burning system, while the inspected parameters were mainly about coking characteristics. This research used measurements to analyze blending fuel characteristics, combustion performance, and pollution, considered necessary for future energy supplies using diversified fuels. In this research, fuel samples with five different HFO blending fuels with asphaltene mass fractions of 4%, 6%, 8%, 16%, and 24% were compared in swirling flame experiments. Pure HFO containing 8% (by mass) asphaltenes was used as a basis for comparison. Low asphaltene content fuel oils were prepared by blending HFO with diesel light oil, and extra asphaltene was added to the HFO to produce high asphaltene oils. This work also intends to illustrate how the mixing of asphaltenes affects the emissions of HFO combustion.

2. Methodology and Fuel Characterization

In general, the burning behavior of different HFOs could lead to the economic development of energy systems with a wide range of fuel adaptability. Heavy crude oil's chemical and physical properties change dramatically from one reservoir to another. Nevertheless, it is currently impossible to test all available HFO samples. In this regard, we recommend that any heavy oil is essentially a mixture of heavy oil, asphaltene, and light oil (diesel). Then, the critical oil characteristics, such as viscosity, density, volatility, asphaltene, and sulfur content, can be adjusted by changing the HFO asphaltene and diesel oil ratio, which determines the combustion characteristics of the oils. In this study, five different HFO samples were prepared by blending HFO with diesel and pure asphaltenes, with asphaltene mass fractions of 4%, 6%, 8%, 16%, and 24%, respectively, given in Table 1. Mixing HFO changes the oil components so that the influence of asphaltene can be thoroughly analyzed. Therefore, this constitutes the main objective of the study. The original HFO sample used in this work featured 8% asphaltene in mass by the test standard IP 143, which was collected from the Shoaiba power plant in Saudi Arabia, and detailed information about these HFO properties can be found in Ref. [30].

Table 1. Properties of the fuel samples used in this study.

Fuel Oil Net Asphaltenes (Mass %)	Composition (Mass %)			Gross Calorific Value (MJ/kg)	H/C Ratio (in Mole)
	Article	Article	Article		
4	50	50	0	45.7	1.68
6	75	25	0	43.4	1.61
8	100	0	0	42.8	1.54
16	91.3	0	8.7	42.5	1.51
24	82.6	0	17.4	42.3	1.47

As shown in Table 2, pure asphaltene was Chinese Standard No. 90 asphaltene, the properties of which are similar to those extracted from HFO [20], and was applied to blending fuel to ensure high asphaltene contents of 16% and 24%. The light oil used was a commercially available diesel to produce low asphaltene content fuels of 4% and 6%. In Figure 1, the asphaltene looked like pitch or soft bitumen at 27 °C, but it became of low liquidity when the temperature rose to 67 °C [4]. Asphaltenes are a class of hydrocarbons with heteroatoms (such as S, N, and O), only defined by their precipitation in nonpolar solvents (such as pentane, hexane, or heptane). Typical asphaltenes include: (a) aromatic rings with alkyl chains up to C30, (b) Sulfur in the benzothiophene ring, (c) Nitrogen in Pyrrole and Pyridine, (d) Porphyrin compounded with vanadium, nickel, and nitrogen-containing pyrrole, (e) Ketones, carboxylic acids, phenols [31].

Table 2. Element compositions of the fuel samples.

			Asphaltene Content in Fuel Samples							
Elements	Pure Asphaltene Elements	Diesel Elements	4%	6%	8%	16%	24%	Unit	LOR	Test Standard
Carbon (C)	84.9	85	85.2	85.3	85.1	84.5	85.2	Mass%	0.1	EPA 440.0
Hydrogen (H)	8.5	13	12.5	11.4	10.9	10.7	10.3	Mass%	0.1	EPA 440.0
Sulfur (S)	5.0	0	1.6	2.3	3.2	3.4	3.6	Mass%	0.1	ASTM D4294
Nitrogen (N)	0.6	0.2	0.2	0.4	0.4	0.4	0.4	Mass%	0.1	EPA 440.0
Metal										
Vanadium (V)	455	-	20.6	39.4	34.2	74.1	82.2	mg/kg	0.2	IP 501–05
Nickel (Ni)	159	-	10.3	14.4	17.8	23.7	35.1	mg/kg	0.5	IP 501–05
Sodium (Na)	273	-	6.5	5.1	7.4	47.5	68.3	mg/kg	2.0	IP 501–05
Magnesium (Mg)	568	-	6.5	<5.0	<5.0	<5.0	46.8	mg/kg	0.1	ASTM D4951

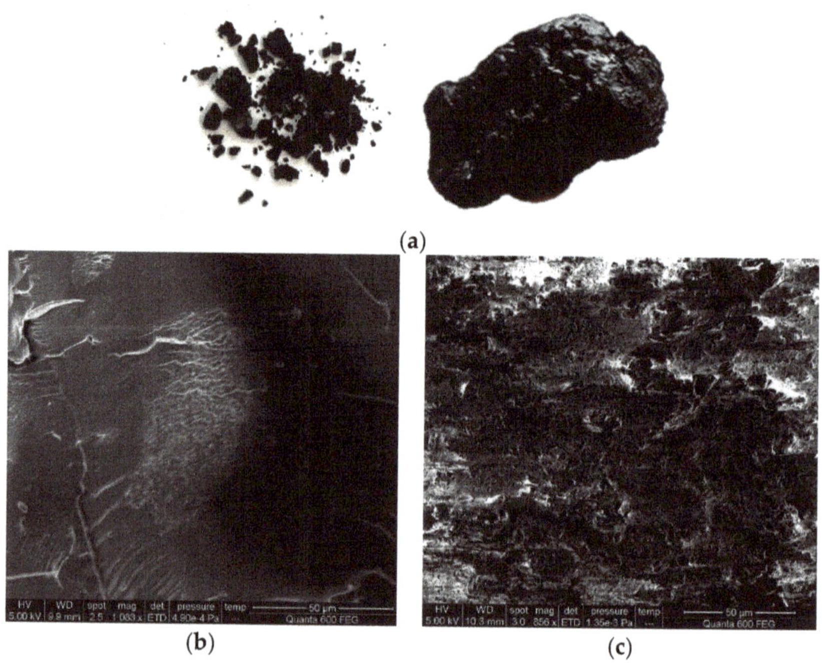

Figure 1. Asphaltene from heavy fuel oil and the morphology (**a**) Asphaltene from heavy fuel oil, (**b**) Surface of asphaltene, (**c**) Cross-section of asphaltene.

Fuel viscosity is critical to fuel spray characteristics and further affects the combustion performance. Viscosity is a measurement of the internal flow resistance of a fluid when it is being deformed. A better understanding of the origin of the high viscosity of heavy oil can help greatly in finding more effective and economical methods to recover heavy oil and reduce the related capital and operating costs. Given this fact, after each blended heavy oil sample was prepared, the viscosity of the fuel was determined by an electromagnetically spinning viscometer (EMS-1000, KEM Kyoto Electronic Manufacturing Co., Ltd., Kyoto, Japan), shown in Figure 2a,b. The viscosity measurement for the fuel sample was carried out three times, and the average value of the three measured viscosity data was noted. The measurement repeatability accuracy (RSD) was 3%. In Figure 2a, the results showed that when the tested temperature ranged from 20.0 °C to 200 °C and the asphaltene fraction ranged from 4 to 24% (wt.%), the heavy oil sample behaved as a Newtonian fluid and its measured viscosity varied in a vast range of 2–568 mPa.s. The viscosities for the five samples decreased exponentially with increasing temperature. In Figure 2a, asphaltene content in heavy oil played a decisive role at high viscosities, especially at low temperatures. Notably, as shown in Figure 2b, for the viscosity of HFOs at 50 °C, the viscosity of fuel oil with 24% asphaltene content was equal to 568 mPa.s, while the viscosity of fuel oil with 4% asphaltene content was only 12 mPa.s at the same temperature, which was almost an order of magnitude lower than that of heavy fuel oil with more than 8% asphaltenes. At a high asphaltene content, the stably attractive interaction between asphaltene particles led to a sharp increase in the viscosity of heavy oil. An increased asphaltene content accompanied the higher viscosity due to the long-range hydrodynamic interactions between maltenes and asphaltene matter [28]. Physically speaking, if solid particles or droplets are so densely dispersed and packed that there is no free space for them to move, the viscosity of the colloidal dispersion is close to infinity [23].

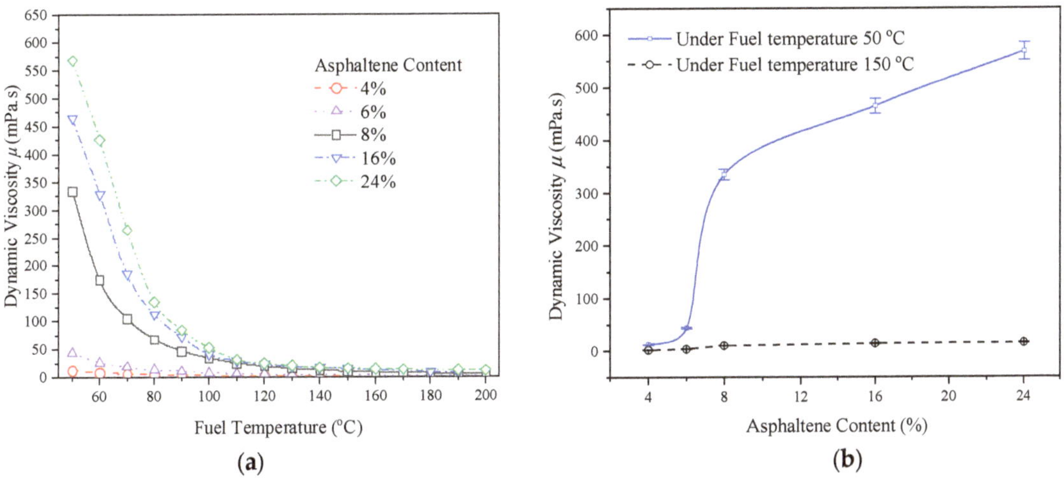

Figure 2. Viscosity distribution of different asphaltene contents of HFO with temperature. (**a**) With temperature variation. (**b**) Viscosity under 50 °C.

The viscosity of heavy oil can be significantly reduced at higher temperatures, especially at high asphaltene contents, as the interparticle interactions among the dispersed asphaltene particles become weak. At a high-temperature range of 150 °C in Figure 2b, the viscosities for the five fuel oils were insensitive to the asphaltene content, as the fuel viscosity came to be relatively low above 100 °C. Since the distance between the asphaltene particles was large enough, the interaction between them could be neglected under relatively high-temperature conditions. In conclusion, asphaltene content determined the high viscosity of heavy oil [25], and these detailed nonlinear reduction results showed that the state of the asphaltene particles in the heavy oil changed with the asphaltene content and temperature, and this change in the state had a significant influence on the viscosity of the heavy oil.

The density of various asphaltene contents heavy oil samples was measured by an electronic density meter (JN-300S/E, Shanghai Jenner Industrial Co., Ltd., Shanghai, China) with a temperature range of 25 °C to 100 °C, as shown in Figure 3. The density of fuel with a high asphaltene content was higher than that for the combination of HFO and light oil. Asphaltenes are polar molecules. Due to their polar properties, they can be used as surfactants to stabilize the interface between the oil molecules in the fuel. Therefore, asphaltenes may result in higher density by tightening the mixture. Under each viscosity, the density basically decreased linearly with the increase of temperature, without a sudden drop process like that of the viscosity. The rate and trend of density decline were essentially constant. For example, as the temperature increased from 25 °C to 100 °C, the density decreased by 7.2% in 4% asphaltene oil and by 7.3% in 24% asphaltene oil. However, as shown in Figure 3b, the density trend was not linear with the variation in asphaltene content. As the asphaltene content increased, the increase in density gradually became slow. For example, the density increased by 8.2% from 4% to 8% asphaltene content fuel, while it only increased by 2% from 8–24% asphaltene cotent fuel at 100 °C.

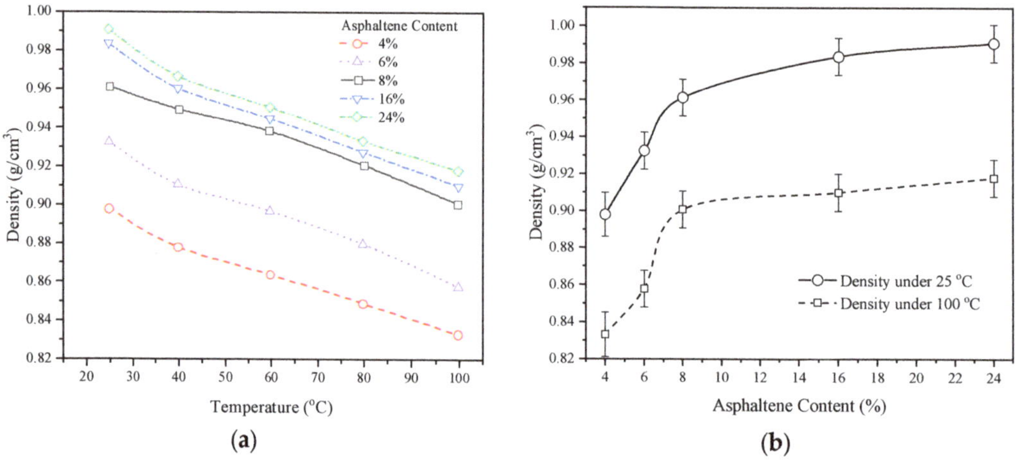

Figure 3. Density distribution of HFO with different asphaltene contents with temperature variation. (**a**) Density with asphaltene content variation. (**b**) Density with temperature variation.

The heating values (gross calorific values) of HFO are reported in Table 1 and were quantified by a Parr Instrument 6400 Automatic Isoperibol calorimeter. The heating values of the HFO samples decreased with increasing asphaltene content, due to the high heating value of light oil (diesel 46 MJ/kg). The highest value was up to 45.7 MJ/kg for the 4% asphaltene fuel, but no predictable linear trend was experienced. The heating value decreased by nearly 8% from 4% to 24% asphaltene fuels. There was a shape decrease trend occurring from 4% to 8% asphaltene fuels with a 6.8% reduction in the heating value, while there was only a 1.2% decrease in the heating values as the fuel asphaltene content increased from 8% to 24%. Notably, regarding the volumetric caloric value, the 24% fuel sample had the highest value of 41,915 MJ/m^3, due to the high-density characteristic of asphaltene.

In Figure 4, the thermogravimetric (TG) and differential thermogravimetric (DTG) curves of five heavy fuel oils under nitrogen and air atmospheres were measured using a Netzsch TG 209 F1Iris thermogravimetric analyzer, with a temperature rise rate of 5 °C/min. In the nitrogen atmosphere, the pyrolysis of HFO is a complex process involving many parallel reactions that coincide, due to the fuel's highly distributed and multicomponent nature. As the fuel is heated, it undergoes devolatilization, including evaporation, distillation, and visbreaking. Due to the absence of oxygen, the HFO molecules decompose and release different hydrocarbon species at different temperatures. As the temperature increases, the rate of mass loss increases, and initially low boiling volatiles are given off, which results in the first peak at approximately 120 °C [17].

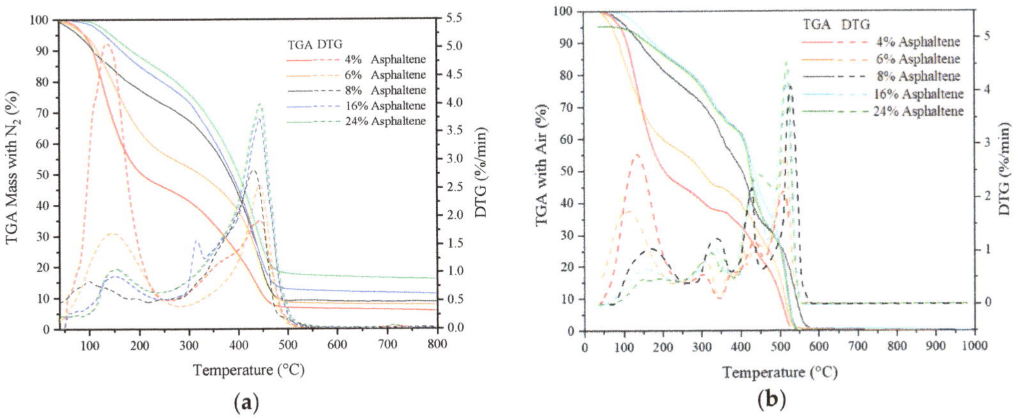

Figure 4. TGA and DTG distribution of HFO with different asphaltene contents. (**a**) N_2 atmosphere. (**b**) Air atmosphere.

The low asphaltene content case of 4% had the highest mass loss rate in the first peak of DTG due to the fuel blended with the light fuel of diesel, in which the rate of light components was much higher than that in pure HFO. The high asphaltene content case of 24% needed a higher temperature to reach the peak of DTG accompanying the lowest mass loss rate in the first peak. With a further increase in temperature up to 250 °C, the rate of mass loss briefly decreased because the bulk of the low boiling volatiles were depleted. There is a second steep increase peaking at 425 °C, due to the release of high boiling volatiles. The high asphaltene cases of 8% and 24% had the same higher mass loss in this peak. Beyond 500 °C, no further apparent mass loss from the fuel was observed. However, the high asphaltene case of 24% had the highest mass left, due to the effect of asphaltene and the high rate of metal compositions in asphaltene. Clearly, the carbon residues were also consistent with the changing trend of asphaltene content. This meant that asphaltene was the origin for carbon formation; thus, the quantity of carbons formed increased with increasing asphaltene content in the oil blend. This result further indicated that the formation of cenospheres was caused by asphaltene during HFO combustion.

In addition, the thermogravimetric (TGA) level in the air condition also corresponded to the asphaltene content in the fuel, shown in Figure 4b. Before the droplet was ignited, the volatiles evaporated considerably. In Figure 4b, under air condition, two peaks were induced by the combustion of low boiling point volatiles and high boiling point matter. These profiles corresponded to those in N_2 with similar temperature regions at approximately 120 °C and 450 °C. It was proven that the fuel components defined the combustion process of HFO in air. The burning of HFO included both the combustion of flammable volatiles and the burn-up of cenospheres and coke residue. Thus, there was no residue after 500 °C in Figure 4b at the end of the test. The typical classified three reaction stages can be identified, in Figure 4b, as LTO (low-temperature oxidation), FD (fuel deposition), and HTO (high-temperature oxidation) [5,32]. As to the pure HFO (8% asphaltene) case, LTO occurred at temperatures below 390 °C with three main processes. Below about 250 °C, the main gaseous products were hydrocarbons with saturated and unsaturated C–H bonds, ascribed to the evaporation of low-boiling point hydrocarbons. Below about 340 °C, oxygen addition reactions occurred with alkyl radicals to generate hydroperoxides; In the temperature range of 340 to 390 °C, the formation of hydroperoxide continued, and the decomposition and isomerization reactions of the hydroperoxides were the dominating reactions releasing carbonyl group, H_2O, CO_2, and CO. In FD between 390 and 460 °C, oxygen-containing compounds produced during the LTO were consumed to produce coke and through cracking or pyrolysis reactions. In HTO, at above 460 °C, the combustion of the coke and solid residues formed in FD process was the dominating reaction. The organic matter was completely oxidized, producing water and carbon oxides along with a production of sulfur and nitrogen oxides depending on the composition of the fuel.

Due to the specific characteristics of asphaltene, heavy fuel oil requires more time and heat to be sufficiently combusted than light distillate oils, such as diesel. Therefore, a swirling flame combustion system was applied to produce a turbulent jet diffusion flame with high vortex stability to burn fuel efficiently [33], shown in Figure 5. The current burner configuration involved relatively simple geometry but replicated many of the fundamental swirling flame behaviors of real gas turbine combustors and boilers.

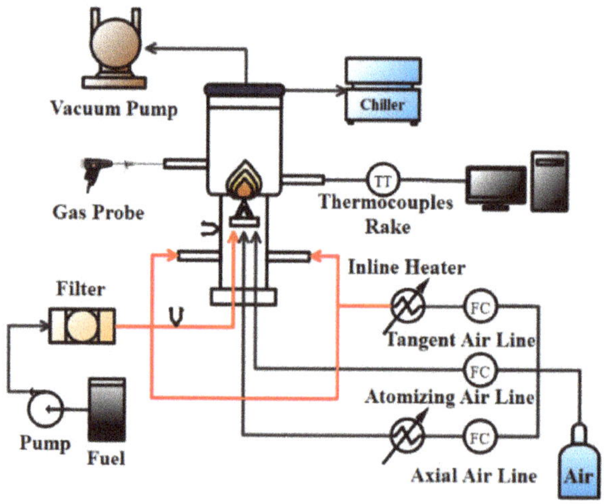

Figure 5. Experimental apparatus.

The swirling flow was generated by a swirl generator with air entering tangentially and axially to cause a swirling air injection, which is referred to the burner of Driscoll [34,35]. The nozzle of Siphon (P/N 30609-2) was used to produce a spray specially designed for viscous liquids. An R-type thermocouple (Pt/13%RhPt, Omega Engineering P13R-010) rake was controlled by a linear translation to measure the radial flame temperature distribution. For more detailed information about this experimental system, please refer to our previous work in Refs. [30,36,37]. In Table 3, the experimental conditions for the five different fuel samples were kept constant and provided with a swirling number of 19.2 calculated by Equation (1):

$$S_g = \frac{(D_a - D_t)D_a}{2D_t^2}\left(\frac{Q_t}{Q_t + Q_a}\right)^2 \tag{1}$$

Table 3. Experimental Conditions.

	Tangent Air (SLPM)	Atomizing Air (SLPM)	Axial Air (SLPM)	HFOs (mL/min)
Flow rate	150	15	5	15
Temperature (°C)	600	150	300	150
Swirling Number S_g		19.2		

The axial and tangential air inlets were applied to control the swirl number. The axial air stream entered through the bottom of the plenum, and its flow rate was Q_a with diameter as D_a. The four tangential air inlets were oriented to the burner tube to impart the swirling momentum with the flow rate represented as Q_t and diameter of D_t. This condition was selected based on the optimization work of the HFO combustion performance evaluation in various swirling flows. We tested the performance of the HFO combustion, under various swirling flame conditions, for different global equivalence ratios, swirling numbers, tangent, and axial airflow rates [37].

3. Results and Discussion

The in-flame temperature distribution of the centerline of the flame of the experiments is given in Figure 6. The uncertainties are given as the standard deviation of the flame temperature profiles. The low asphaltene content case of 4% had the highest temperature distribution, ascribed to complete combustion performance due to the fine spray characteristic (low viscosity and density) and high rate of the light components (in Figure 4) of the fuel. In addition, the heating value of diesel (46 MJ/kg) was slightly higher than that of pure HFO (42.4 MJ/kg). The high H level in 4% asphaltene content in Table 1 also increased the temperature. The fuels with 6% and 8% asphaltene contents exhibited almost the same level of temperature profiles, due to the similar fuel properties introduced in the fuel characteristics investigation above, in Table 2. The fuel sample with 16% asphaltene was in the state between 8% and 24% asphaltene conditions, which could be regarded as the critical condition for sufficient combustion. The results of the gaseous and solid particle emission in the next sections further confirmed that the 16% asphaltene was a critical value for the acceptable concertation of high asphaltene fuel combustion in the range of asphaltene contents in this study.

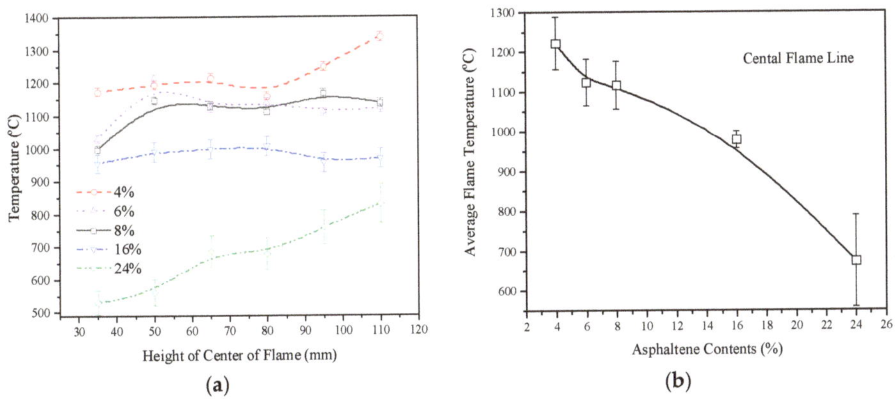

Figure 6. In-flame temperature measurements on the centerline of the flame. (**a**) Center flame temperature profiles. (**b**) Average flame temperature.

The 24% asphaltene fuel oil had an obviously low-temperature distribution in comparison to the other cases. Asphaltene is the main factor that induces incomplete combustion. In the observation of the experiment of 24% asphaltene HFO, the fuel was very difficult to combust, and the flame was very unstable during the experiment, which meant that the heat could not accumulate well. The incomplete combustion of the high asphaltene experiment could be ascribed to the following factors: (a) The density of the asphaltene was higher than that of pure HFO, and the density of the high asphaltene fuel was higher than that of pure HFO. In our experiment, the flow rates were constant in volume at 15 mL/min, and more fuel entered the burner in the high asphaltene case, which brought about changes in the global equivalence ratio, approaching the fuel-rich condition. (b) The high viscosity of 24% asphaltene fuel inducing low-quality spray flow also made the combustion incomplete. (c) According to the results from our previous investigation on suspended droplet experiments, the ignition delay time of high asphaltene content was higher than that of low asphaltene fuel under the same droplet size [5]. Therefore, the high asphaltene fuel droplets needed more time to be combusted.

In Figure 6a, before the droplet was ignited, the volatiles in the fuel needed to evaporate first by absorbing a large amount of heat. These factors resulted in the relatively low temperature location in the root of the flame, especially for the 24% asphaltene case. In Figure 6b, the average measurement flame temperature of fuel oils was subject to a linear relationship with respect to the asphaltene content. This research might point to a viable way to burn these residues by mixing them with low-viscosity and high-volatility fuels. In this case, the burning characteristics of the blending fuels introduced in this work are essential for technology development.

The flue gas emissions of the five different asphaltene content swirling flame experiments are given in Figure 7. When a stable HFO swirl flame was obtained, a gas analyzer (TESTO 350) was used to measure the gaseous emissions of CO, SO_2, NO_x, O_2 and CO_2 from the top of the furnace in the overfire area, a method which has been widely applied to the studies of pollutants measurement of combustion devices [38–40]. Each component's gas emission measurement result was averaged from the data within 10 min and normalized to 15% oxygen applying Equation (2) to eliminate the influence of the dilution air as:

$$C[@15\%O_2] = C_m\ (20.9 - 15)/(20.9 - O_2\%) \quad (2)$$

where $C[@15\%O_2]$ was the calculated concertation eliminating the dilution air effect, C_m was the gaseous pollution measurement content [41]. The detailed measurement method description, validation and uncertainties can be found in our previous work [30,37]. HFO is characterized by a high asphaltene content (approximately 8% by mass), which leads to incomplete combustion. Asphaltenes are also responsible for the high metal content of fuels because of the presence of embedded heteroatoms. In Figure 7a, the general trend of the CO and CO_2 profiles rose with increasing asphaltene content. The Modified Combustion Efficiency (MCE) was also calculated by $CO_2/(CO + CO_2)$ to determine the combustion efficiency at different asphaltene concentrations. The 24% asphaltene case had a high level of oxygen (up to 13%) left which proved the incomplete combustion of high asphaltene content experiment compared to the low asphaltene cases, which also resulted in the high-level emission of CO. The CO emissions of 4% and 6% asphaltene case were higher than that of the 8% asphaltene case, due to the higher carbon concentration in the 4% asphaltene fuel sample, as shown in Table 1.

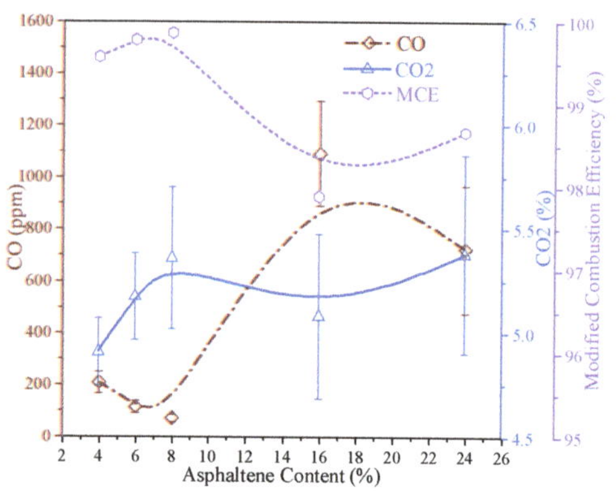

(a)

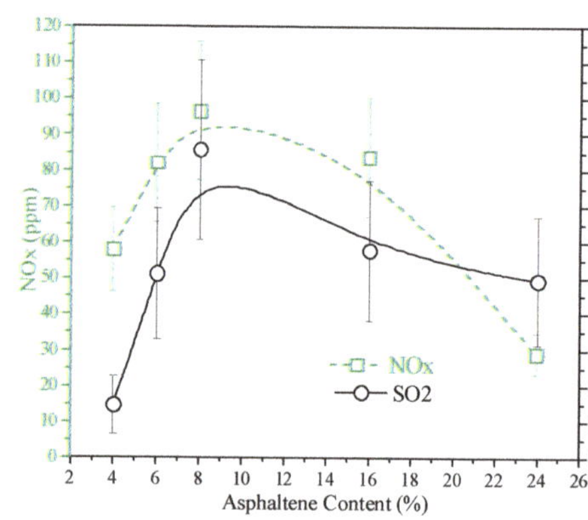

(b)

Figure 7. Gaseous emissions of fuel combustion with different asphaltene contents (**a**) CO and CO_2. (**b**) NO_x and SO_2.

NO_x refers to oxides of nitrogen and can induce smog formation, acid rain, and ozone depletion in the upper atmosphere. In Figure 7b, the trend of NO_x decreased with the increase in asphaltene content in the fuel, due to the declined average temperature profiles, in Figure 6a, having a suppression effect on the combustion temperature of high asphaltene fuel. The formation of NO_x can be classified into three mechanisms: thermal, prompt, and fuel thermal. Thermal NO_x is the dominant mechanism in combustion processes above 1100 °C, which is common in most high-temperature heating applications [42]. In Figure 6b, the average flame temperature of low-asphaltene heavy oil (4%, 6%, 8%) combustion was above 1100 °C, and its maximum temperature could reach 1125 °C, while the average temperatures of the other two cases were below 1100 °C. Prompt NO_x usually predominates in fuel-rich conditions formed by relatively fast reactions between nitrogen, oxygen, and hydrocarbon radicals. In our experimental study, the global equivalence ratio of pure HFO (8% asphaltenes) was 0.91. Therefore, the combustion of low-asphaltene fuels (4% and 6%) occurred under fuel-lean conditions, while the combustion of high-asphaltene fuels was under fuel-rich conditions. Fuel NO_x comes from the direct oxidation reaction of organic-nitrogen compounds in the fuel. HFO contains a large amount of bound nitrogen that increased as the asphaltene content increased, as shown in Table 1, with the maximum value up to 0.431% in mass for the 24% case. The conversion of fuel-bound nitrogen to NO_x was between 15% and 100%. In general, the higher the conversion efficiency is, the lower the nitrogen content in the fuel. Therefore, in this study, the mechanism of NO_x formation varied with the fuel asphaltene content. This was the reason for the high level of NO_x emissions from 6% and 8% asphaltene fuel combustion.

In Figure 7b, the trend of SO_2 slightly increased with increasing asphaltene content in the fuel. Additionally, SO_x reactions may inhibit fuel NO_x formation. The low SO_2 emission in the 4% asphaltene experiment was caused by the low concentration of sulfur in the fuel sample, as shown in Table 1. The incomplete combustion of high asphaltene fuel of 24% was the reason for the low SO_2 emission of the fuel, which also showed that some residual fuel attached to the filter, as shown in Figure 8e. Sulfur oxide emission is a critical problem with HFO combustion. The primary source of SO_x comes from HFOs that contain the highest sulfur, with 3.36% in 24% asphaltene fuel, as shown in Table 1. Notably, in the five oil samples, the sulfur content in the fuel increased nonlinearly with the increase in asphaltenes. The sulfur content decreased significantly after adding light oil, but by adding pure asphaltene, the effect of increasing sulfur content was not pronounced, which indicated that the source of sulfur in the heavy fuel oil was not all asphaltene. Sulfur dioxide (SO_2) tends to be the preferential product at higher temperatures, while sulfur trioxide (SO_3) is more likely to form at lower temperatures. Since most combustion processes are at high temperatures, SO_2 is the most common form of SO_x emitted from systems containing sulfur. According to the sulfur balance calculation, most sulfur comes to flue gas emissions forming SO_2 and SO_3, and the remaining sulfur goes into the PM.

A comparison of the morphologies of five asphaltene samples is provided in Figure 8. The cenospheres were obviously distinguished in the SEM measurement results. The general characteristics of burning heavy oil droplets can be clearly identified, such as micro-explosions, the formation of smoke and coke, and coke residue smoldering. Note that combustion in the solid phase is actually carried out without flame generation [4]. For coke formation, the cenosphere is where the evolution of volatile substances abruptly ends, and the droplets collapse to form hard carbonaceous residues. Oils with 6%, 8%, and 16% asphaltene morphology showed minimal shrinkage and formed large thin-walled coke shells, similar in size to the original spray droplets.

In general, compared to the 8% asphaltene sample, the 4% asphaltene sample had an accumulation distribution that was more powdery and more small particles but fewer cenospheres, which could be ascribed to the small spray droplet size, fine spray, and good combustion performance. In Figure 8e, the SEM result of 24% asphaltene illustrated the incomplete combustion of the high asphaltene experiment in which the cenosphere and filter fiber were all covered by unburned fuel. The relatively long ignition delay time of the high asphaltene sample made the fuel absorb heat and vapor but not combust, and further attached to the filter due to the cooling device above the filter.

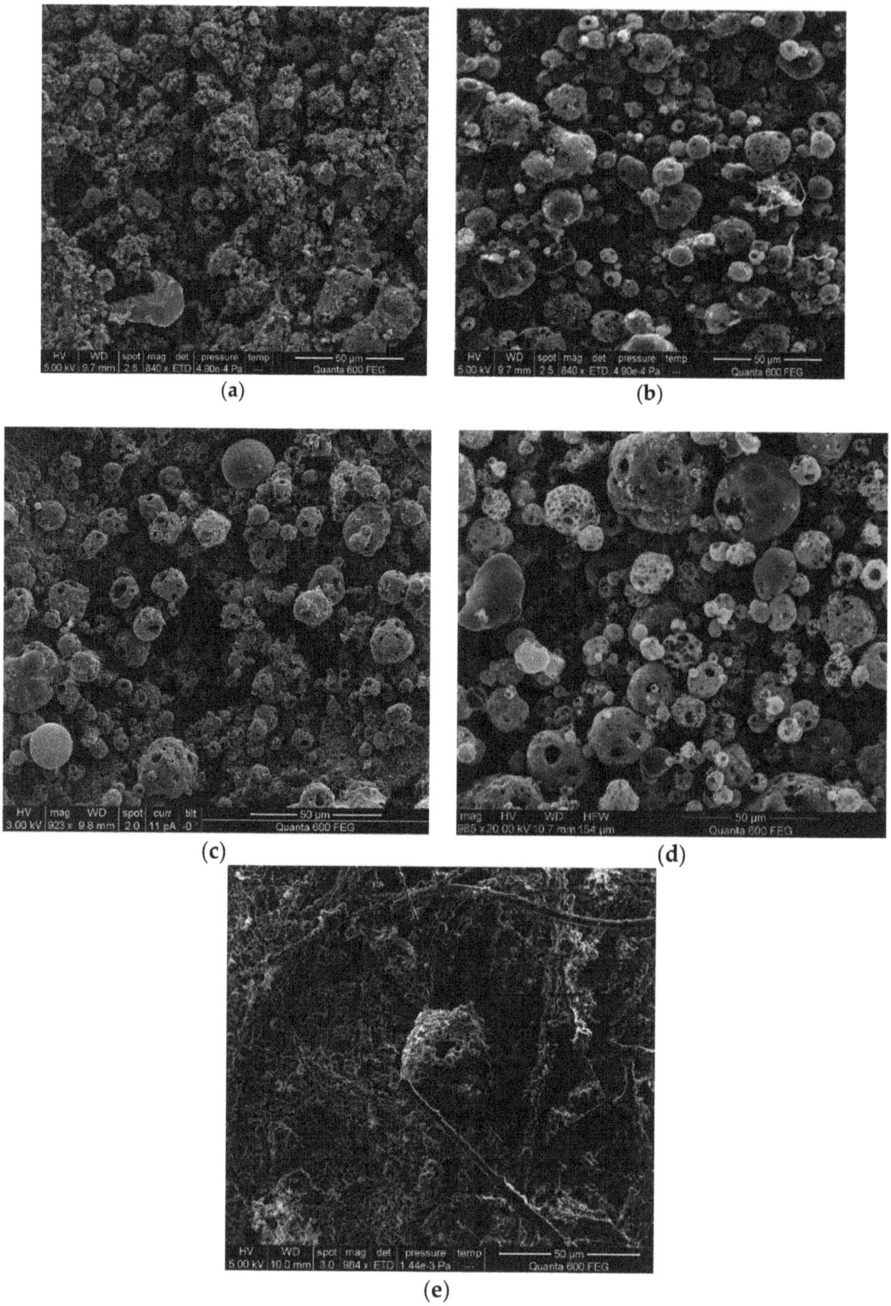

Figure 8. Morphology of the solid particle emissions from fuel combustion with various asphaltene contents. (**a**) 4% asphaltene HFO. (**b**) 6% asphaltene HFO. (**c**) 8% asphaltene HFO. (**d**) 16% asphaltene HFO. (**e**) 24% asphaltene HFO.

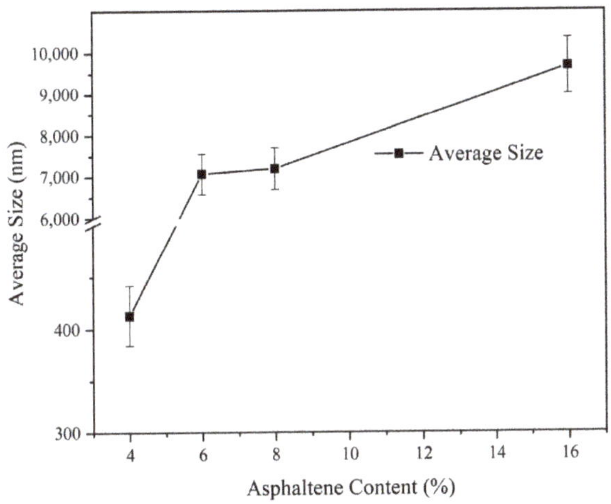

Figure 9. Average size of the solid particle emission.

The averaged sizes of solid particles collected from the filter of the four experiments measured by the Zetasizer Nano ZS90 are compared in Figure 9, except for the 24% one because the particles of 24% fuel were all stuck on the filter and were hard to collect. Based on the morphology of the 24% sample, we could not ascribe the increase in the deposition rate in the particle emissions because a lot of fuel attached to the filter. For this reason, we could not scrape the solid deposition on the filter from the sample of 24% asphaltene content fuel, and it was also impossible to analyze the composition by ICP/ODS. The average size of the particle increased with the increase in the asphaltene content of fuel, due to a strong correspondence between the spray droplet size and the diameter of the cenosphere [30]. The size of 16% asphaltene fuel was nearly 23 times larger than that of the 4% asphaltene case. The deposition rates are given in Figure 10. The low asphaltene experiment of 4% had the lowest deposition rate, which could be ascribed to fine combustion and the low metal compositions of a low level in the 4% asphaltene fuel sample.

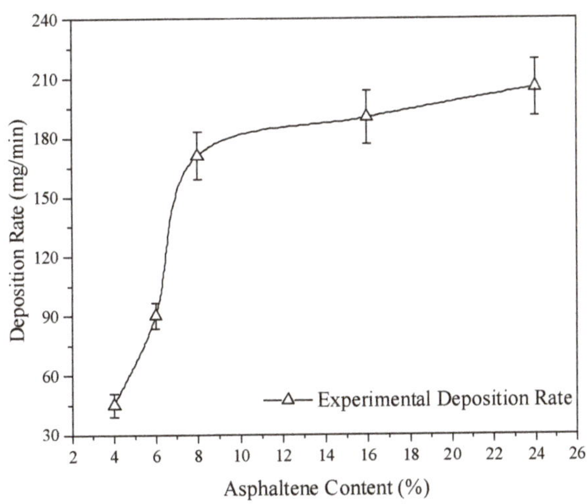

Figure 10. Comparison of the deposition rate of HFO with different asphaltene contents.

The components of solid particle emissions from the representative cases of 4%, 8%, and 16% asphaltene experiments are compared in Table 4. The sulfur content in solid particulates increased with increasing asphaltene content in the fuel. The high level of metal components in the solid particle emissions corresponded to the composition of pure asphaltene. High levels of sulfur and vanadium induced cold corrosion and hot corrosion for burners, which are unacceptable for fuel applications. Thus, asphaltene caused the formation of cenospheres. The TG diagrams of the blended oil combustion particles are compared in Figure 11. Unlike the TGA results of the fuel in air, the TG results of the particles after combustion were similar, indicating fine combustion performance with these three asphaltene fuels.

Table 4. Analysis of the compositions of solid particle emissions.

Elements	Units	4%	8%	16%	LOR	Test Standard
Carbon (C)	Mass%	77	81	80	0.1	EPA 440.0
Sulfur (S)	Mass%	2.6	3.9	5.4	0.1	ASTM D4294
Nitrogen (N)	Mass%	0.8	0.6	0.7	0.1	EPA 440.0
Oxygen (O)	Mass%	18	10	5	5.0	EPA 440.0
Metal Elements						
Vanadium (V)	mg/kg	802	798	3312	0.2	IP 501–05

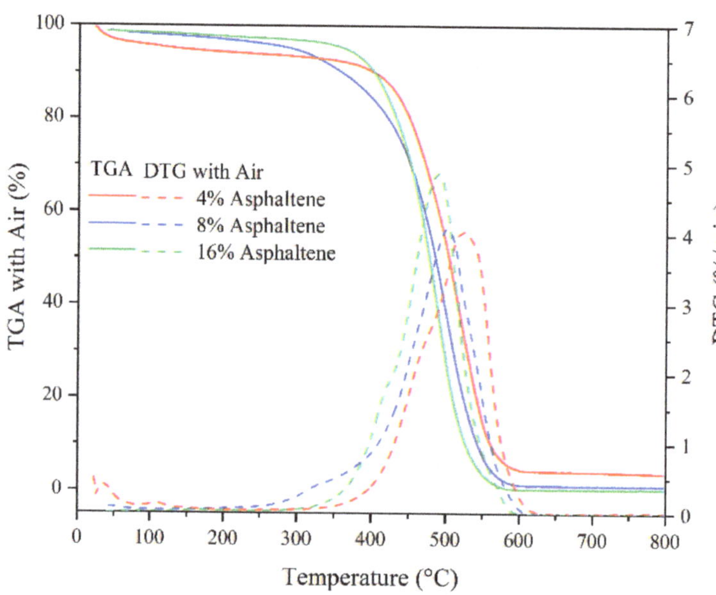

Figure 11. TGA and DTG distribution of solid particle emissions.

4. Conclusions

This study introduces an experimental study on the specific effects of asphaltene content on the fuel properties and swirling flame combustion performance of heavy oil fuels under five different asphaltene contents of 4%, 6%, 8%, 16%, and 24%. The analysis of the properties of fuel samples with various asphaltene contents, including composition, density, viscosity, heating value, and TG analysis, showed the importance of all fuel properties that must be considered prior to implementation and combustion. The asphaltene content determines the high viscosity of heavy oil. The viscosity of heavy oil can be significantly reduced at higher temperatures, especially at high asphaltene contents, as the interparticle interactions among the dispersed asphaltene particles become weaker. The nonlinear viscosity reduction result revealed that the state of the asphaltene particles in the heavy oil varies with asphaltene concentration and temperature, and this change in state has a significant impact on the viscosity of the heavy oil. Since blended light oil has a high heating value, the heating values of the HFO samples drop as the asphaltene percentage increases. The TGA and morphological studies showed that asphaltene is vital for cenosphere formation during HFO combustion. The reduction in asphaltene percentage in HFO enhances spray characteristics and combustion performance, as well as lowers particle emissions, including those of sulfur and metal components, such as vanadium, avoiding hot and cold corrosion on the burner's surface. The mechanism of NO_x formation varies with the fuel asphaltene content, due to the different combustion performances. The 4% asphaltene sample had an accumulation distribution that was more powdery and more tiny particles but less cenospheres than the 8% asphaltene sample, attributed to the small spray droplet size, fine spray, and efficient combustion. The high amount of metal components in solid particle emissions corresponds to pure asphaltene compositions. As a result, asphaltene contributes to the formation of cenospheres. In general, the combustion emissions of various asphaltene fuels are summarized by investigating the impact of asphaltenes on the fuel properties and burning behavior of HFO blends. The research may indicate a potential method of burning HFO by combining it with low-viscosity and high-volatility fuels. In this instance, the combustion properties of the blending fuels proposed in this study are critical for technological advancement, and the further achieving of low carbon emission targets.

Author Contributions: Conceptualization, X.P. and W.L.R.; methodology, X.P. and H.T.; formal analysis, X.P.; investigation, X.P.; writing—review and editing, X.P. and H.T.; funding acquisition, X.P. and W.L.R. All authors have read and agreed to the published version of the manuscript.

Funding: This research was funded by National Natural Science Foundation of China (grant number 52006122) and National Science and Technology Major Project (grant number 2019-III-0014-0058).

Institutional Review Board Statement: Not applicable.

Informed Consent Statement: Not applicable.

Data Availability Statement: Not applicable.

Acknowledgments: The research was in collaboration with the Clean Combustion Research Center (CCRC) at King Abdullah University of Science and Technology (KAUST).

Conflicts of Interest: The authors declare no conflict of interest.

References

1. Czernik, S.; Bridgwater, A.V. Overview of Applications of Biomass Fast Pyrolysis Oil. *Energy Fuels* **2004**, *18*, 590–598. [CrossRef]
2. Ilyin, S.; Arinina, M.; Polyakova, M.; Bondarenko, G.; Konstantinov, I.; Kulichikhin, V.; Malkin, A. Asphaltenes in heavy crude oil: Designation, precipitation, solutions, and effects on viscosity. *J. Pet. Sci. Eng.* **2016**, *147*, 211–217. [CrossRef]
3. Pereira, T.M.C.; Vanini, G.; Oliveira, E.C.S.; Cardoso, F.M.R.; Fleming, F.P.; Neto, A.C.; Lacerda, V.; Castro, E.V.R.; Vaz, B.G.; Romão, W. An evaluation of the aromaticity of asphaltenes using atmospheric pressure photoionization Fourier transform ion cyclotron resonance mass spectrometry—APPI(±)FT-ICR MS. *Fuel* **2014**, *118*, 348–357. [CrossRef]
4. Xu, G.; Ikegami, M.; Honma, S.; Ikeda, K.; Nagaishi, H.; Takeshita, Y. Burning droplets of heavy oil residual blended with diesel light oil: Characterization of burning steps. *Combust. Sci. Technol.* **2002**, *174*, 115–145. [CrossRef]
5. Khateeb, A.; Elbaz, A.M.; Guida, P.; Roberts, W.L. Influence of asphaltene concentration on the combustion of a heavy fuel oil droplet. *Energy Fuels* **2018**, *32*, 12981–12991. [CrossRef]
6. Villasenor, R.; Garcia, F. An experimental study of the effects of asphaltenes on heavy fuel oil droplet combustion. *Fuel* **1999**, *78*, 933–944. [CrossRef]
7. Elbaz, A.M.; Roberts, W.L. PM from the combustion of heavy fuel oils. *Energy* **2018**, *152*, 455–465. [CrossRef]
8. Marrone, N.J.; Kennedy, I.M.; Dryer, F.L. Coke formation in the combustion of isolated heavy oil droplets. *Combust. Sci. Technol.* **1984**, *36*, 149–170. [CrossRef]
9. Jiang, L.; Elbaz, A.M.; Guida, P.; Al-Noman, S.M.; AlGhamdi, I.A.; Saxena, S.; Roberts, W.L. Cenosphere formation during single-droplet combustion of heavy fuel oil. *Energy Fuels* **2019**, *33*, 1570–1581. [CrossRef]

10. Moszkowicz, P.; Witzel, L.; Claus, G. Modelling of very fast pyrolysis of heavy fuel oil droplets. *Chem. Eng. Sci.* **1996**, *51*, 4075–4086. [CrossRef]
11. Azimi, A.; Arabkhalaj, A.; Markadeh, R.S.; Ghassemi, H. Fully transient modeling of the heavy fuel oil droplets evaporation. *Fuel* **2018**, *230*, 52–63. [CrossRef]
12. Fostiropoulos, S.; Strotos, G.; Nikolopoulos, N.; Gavaises, M. Numerical investigation of heavy fuel oil droplet breakup enhancement with water emulsions. *Fuel* **2020**, *278*, 118381. [CrossRef]
13. Garaniya, V.; Goldsworthy, L. Vaporization and pyrolysis modelling of a single droplet of heavy fuel oil using continuous thermodynamics. *ANZIAM J.* **2007**, *49*, C116–C138. [CrossRef]
14. Kwack, E.; Shakkottai, P.; Massier, P.; Back, L. Morphology of globules and cenospheres in heavy fuel oil burner experiments. *J. Eng. Gas Turbines Power* **1992**, *114*, 338–345. [CrossRef]
15. Ambalae, A.; Mahinpey, N.; Freitag, N. Thermogravimetric studies on pyrolysis and combustion behavior of a heavy oil and its asphaltenes. *Energy Fuels* **2006**, *20*, 560–565. [CrossRef]
16. Atiku, F.A.; Bartle, K.D.; Jones, J.M.; Lea-Langton, A.R.; Williams, A. A study of the combustion chemistry of petroleum and bio-fuel oil asphaltenes. *Fuel* **2016**, *182*, 517–524. [CrossRef]
17. Abdul Jameel, A.G.; Han, Y.; Brignoli, O.; Telalović, S.; Elbaz, A.M.; Im, H.G.; Roberts, W.L.; Sarathy, S.M. Heavy fuel oil pyrolysis and combustion: Kinetics and evolved gases investigated by TGA-FTIR. *J. Anal. Appl. Pyrolysis* **2017**, *127*, 183–195. [CrossRef]
18. Cunningham, A.T.S.; Gliddon, B.J.; Jackson, P.J.; Jones, A.R.; Lawn, C.J.; Sarjeant, M.; Squires, R.T.; Street, P.J. The esso energy award lecture, 1988—Improvements in the combustion of heavy fuel oils. *Proc. R. Soc. Lond. Ser. A Math. Phys. Eng. Sci.* **1989**, *423*, 233–265.
19. Urban, D.L.; Huey, S.; Dryer, F.L. Evaluation of the coke formation potential of residual fuel oils. *Symp. Combust.* **1992**, *24*, 1357–1364. [CrossRef]
20. Goldstein, H.L.; Siegmund, C.W. Influence of heavy fuel oil composition and boiler combustion conditions on particulate emissions. *Environ. Sci. Technol.* **1976**, *10*, 1109–1114. [CrossRef]
21. Geoffrey, C.; Allen, W.H.; Hocking, D.G.; Watson, R.; Wild, P.J. Street, Surface studies of coke particles from residual fuel oil combustion. *J. Inst. Energy* **1984**, *57*, 260–265.
22. Walsh, P.M.; Olen, K.R. Emission of unburned coke from combustion of residual fuel-oil in wall-fired electric utility boilers. *J. Inst. Energy* **1993**, *66*, 140–146.
23. Gu, Y. Effects of asphaltene content on the heavy oil viscosity at different temperatures. *Fuel* **2007**, *86*, 1069–1078.
24. Munro, A.; Westlake, D.; Lewis, A. The conservative use of liquid fuel. *J. Inst. Fuel* **1978**, *51*, 10–19.
25. Sheu, E.Y.; Mullins, O.C. *Asphaltenes: Fundamentals and Applications*; Springer: Berlin, Germany, 1995.
26. Fakher, S.; Ahdaya, M.; Elturki, M.; Imqam, A. Critical review of asphaltene properties and factors impacting its stability in crude oil. *J. Pet. Explor. Prod. Technol.* **2020**, *10*, 1183–1200. [CrossRef]
27. Schuler, B.; Meyer, G.; Pena, D.; Mullins, O.C.; Gross, L. Unraveling the molecular structures of asphaltenes by atomic force microscopy. *J. Am. Chem. Soc.* **2015**, *137*, 9870–9876. [CrossRef] [PubMed]
28. Peng, L.; Gu, Y. Effects of asphaltene content and solvent concentration on heavy oil viscosity. In Proceedings of the SPE International Thermal Operations & Heavy Oil Symposium, Calgary, AB, Canada, 1–3 November 2005.
29. Bartle, K.; Jones, J.; Lea-Langton, A.; Pourkashanian, M.; Ross, A.; Thillaimuthu, J.; Waller, P.; Williams, A. The combustion of droplets of high-asphaltene heavy oils. *Fuel* **2013**, *103*, 835–842. [CrossRef]
30. Pei, X.; Guida, P.; AlAhmadi, K.M.; Al Ghamdi, I.A.; Saxena, S.; Roberts, W.L. Cenosphere formation of heavy fuel oil/water emulsion combustion in a swirling flame. *Fuel Process. Technol.* **2021**, *216*, 106800. [CrossRef]
31. Zheng, F.; Shi, Q.; Vallverdu, G.S.; Giusti, P.; Bouyssiere, B. Fractionation and characterization of petroleum asphaltene: Focus on metalopetroleomics. *Processes* **2020**, *8*, 1504. [CrossRef]
32. Yuan, C.; Emelianov, D.A.; Varfolomeev, M.A.; Rodionov, N.O.; Suwaid, M.A.; Vakhitov, I.R. Mechanistic and kinetic insight into catalytic oxidation process of heavy oil in in-situ combustion process using copper (II) stearate as oil soluble catalyst. *Fuel* **2021**, *284*, 118981. [CrossRef]
33. Bohon, M.D. Experimental and Kinetic Investigation of the Influence of OH Groups on NO_x Formation. Ph.D. Thesis, King Abdullah University of Science and Technology, Thuwal, Saudi Arabia, 2016.
34. Feikema, D.; Chen, R.-H.; Driscoll, J.F. Enhancement of flame blowout limits by the use of swirl. *Combust. Flame* **1990**, *80*, 183–195. [CrossRef]
35. Chen, R.-H.; Driscoll, J.F.; Kelly, J.; Namazian, M.; Schefer, R. A comparison of bluff-body and swirl-stabilized flames. *Combust. Sci. Technol.* **1990**, *71*, 197–217. [CrossRef]
36. Pei, X.; Jameel, A.G.A.; Chen, C.; Alghamdi, I.A.; Roberts, W.L. Swirling flame combustion of heavy fuel oil: Effect of fuel sulfur content. *J. Energy Resour. Technol.* **2021**, *143*, 082103. [CrossRef]
37. Pei, X.; Elbaz, A.M.; Jiang, L.; AlAhmadi, K.M.; Saxena, S.; Roberts, W.L. Heavy fuel oil combustion characteristics evaluation in various swirling flow conditions. *J. Eng. Gas Turbines Power-Trans. ASME* **2021**, *143*, 071012. [CrossRef]
38. Ardebili, S.M.S.; Kocakulak, T.; Aytav, E.; Calam, A. Investigation of the effect of JP-8 fuel and biodiesel fuel mixture on engine performance and emissions by experimental and statistical methods. *Energy* **2022**, *254*, 124155. [CrossRef]

39. Ithnin, A.M.; Ahmad, M.A.; Bakar, M.A.A.; Rajoo, S.; Yahya, W.J. Combustion performance and emission analysis of diesel engine fuelled with water-in-diesel emulsion fuel made from low-grade diesel fuel. *Energy Convers. Manag.* **2015**, *90*, 375–382. [CrossRef]
40. Yang, N.; Deng, X.; Liu, B.; Li, L.; Li, Y.; Li, P.; Tang, M.; Wu, L. Combustion performance and emission characteristics of marine engine burning with different biodiesel. *Energies* **2022**, *15*, 5177. [CrossRef]
41. Baukal, C.E., Jr. *The John Zink Hamworthy Combustion Handbook: Volume 1—Fundamentals*; CRC Press: Abingdon, UK, 2012.
42. Hoda, S.N.; Nassab, S.A.G.; Ebrahim, J.J. Three dimensional numerical simulation of combustion and heat transfer in porous radiant burners. *Int. J. Therm. Sci.* **2019**, *145*, 106024. [CrossRef]

energies

MDPI

Article

Composition of Flue Gases during Oxy-Combustion of Energy Crops in a Circulating Fluidized Bed

Monika Kosowska-Golachowska [1], Adam Luckos [2] and Tomasz Czakiert [3,*]

1 Department of Thermal Machinery, Czestochowa University of Technology, Armii Krajowej 21, 42-201 Czestochowa, Poland
2 Braamfontein Campus East, School of Chemical and Metallurgical Engineering, University of the Witwatersrand, Johannesburg 2050, South Africa
3 Department of Advanced Energy Technologies, Czestochowa University of Technology, Dabrowskiego 73, 42-201 Czestochowa, Poland
* Correspondence: tczakiert@is.pcz.pl

Abstract: In recent years, global warming and climate change associated with emissions of CO_2 from fossil fuel-fired power systems are a big worry for authorities in many countries worldwide. The utilization of biomass as an alternative, carbon-neutral fuel can reduce emissions of CO_2 and other greenhouse gases. Furthermore, the coupling of oxy-combustion of biomass with CO_2 capture is an option for carbon-negative power generation technology. In this study, emissions of NO_x, SO_2, and CO from the air- and oxy-combustion of three energy crops (*Miscanthus giganteus*, *Sida hermaphrodita*, and *Salix viminalis*) are presented and compared with emissions from other biomass fuels and reference coal. Combustion tests in air and O_2/CO_2 mixtures were conducted in a 12-kW bench-scale CFB combustor at 850 °C. Measurements of flue gas compositions were taken using an FTIR spectrometer. In all tested atmospheres, emissions of SO_2, N_2O, and CO for biomass were lower than those for the reference coal. The oxidation of volatile nitrogen compounds was behind high emissions of NO_x from biomass burned in air and O_2/CO_2 mixtures. The lowest concentrations of NO were found in the 21% O_2/70% CO_2 mixture. Combustion in mixtures containing more oxygen (30% and 40% O_2) led to a decrease in emissions of N_2O and CO and an increase in emissions of NO and SO_2.

Keywords: biomass; oxy-fuel combustion; fluidized bed; *Miscanthus giganteus*; *Sida hermaphrodita*; *Salix viminalis*; wheat straw; Scots pine; bituminous coal

Citation: Kosowska-Golachowska, M.; Luckos, A.; Czakiert, T. Composition of Flue Gases during Oxy-Combustion of Energy Crops in a Circulating Fluidized Bed. *Energies* **2022**, *15*, 6889. https://doi.org/10.3390/en15196889

Academic Editor: Albert Ratner

Received: 23 August 2022
Accepted: 16 September 2022
Published: 20 September 2022

1. Introduction

Various human activities related to the use of fossil fuels such as electricity generation, agriculture, and transport release large quantities of CO_2 into the atmosphere. Carbon dioxide as a greenhouse gas (GHG) is a major contributor to global warming. In the transition to a low-carbon economy, i.e., in the next few decades, the world will still be heavily dependent on fossil fuels as primary energy sources despite unprecedented progress in solar and wind energy generation. Therefore, the utilization of advanced systems for CO_2 capture, utilization, and sequestration (CCUS) becomes vital to abate emissions and lessen the impact of CO_2 emissions on climate change. Coupling CCUS with renewables should help to reach the 2 °C temperature rise limit recommended by the Paris Agreement.

The contribution of biomass to global primary energy consumption is the largest among all sources of renewable energy [1]. Biomass is carbon neutral because, unlike fossil fuels, it is a part of the natural carbon cycle. Carbon dioxide released during the combustion of biomass is absorbed by plants during their life cycles.

The basic characteristics of biomass as a fuel include moisture content, ash yield and composition, carbon content, and higher heating value (HHV). Vassilev and co-workers [2] reviewed the chemical composition of many biomass crops and found large variability in moisture content, ash yield, and types of inorganic matter (minerals) in raw materials. However, the results of proximate and ultimate analyses when reported on a daf (dry, ash-free) basis were in a narrow range.

Among various biomass sources, dedicated energy crops (non-food crops, mostly perennial grasses, and fast-growing trees) are grown solely to provide renewable feedstock for the generation of electricity and heat and the production of biofuels [3]. These high biomass yield crops can be cultivated on marginal or reclaimed land; they are non-invasive and have low water and fertilizer needs. Herbaceous energy crops, e.g., miscanthus and switchgrass, can store approximately two times more CO_2 per 1 hectare of cultivation than woody crops such as poplar and willow.

Among CCUS technologies, oxy-fuel combustion is a promising process because it uses almost pure oxygen instead of air and generates a high-purity stream of CO_2 (95% or more) that can be stored or utilized to produce carbon-based fuels or chemicals. Biomass firing or co-firing in oxy-fuel systems integrated with CO_2 capture can be considered a carbon-negative technology.

Fluidized bed boilers, both stationary and circulating, are the best choice for firing low-quality coals, refuse-derived fuels, and biomass [4–6]. The well-known benefits of fluidized-bed combustion technology include high fuel flexibility and turndown ratio, low emissions of NO_x, and high availability. Today, circulating fluidized-bed (CFB) boilers firing 100% biomass fuels are available commercially. Units at Konin (55 MW_e) and Połaniec (205 MW_e) power plants in Poland [7], and the 299 MW_e Tees unit in the UK (the largest bio CFB in the world) [8], can be mentioned as examples.

Oxy-fuel CFB boilers are better than their pulverized-fuel (PF) counterparts. They do not require sophisticated burners and fuel preparation and feeding systems. Because the combustion temperature in oxy-CFB systems is lower than in PF units, oxy-CFB boilers can meet current NO_x and SO_x emission limits without additional de-NO_x and de-SO_x equipment. The combustion temperature can be controlled by cooling a part of the circulating particles in an external heat exchanger and recirculating them back to the combustion chamber [9]. This allows for using higher initial O_2 concentrations in smaller streams of recirculated flue gas, which results in units with reduced dimensions, higher thermal efficiency, and lower capital costs. Energy losses in oxy-fuel systems caused by air separation and CO_2 purification systems working at high pressures can be reduced in oxy-CFB units operating at elevated pressure. The latent heat in flue gases can be recovered and air infiltration can be eliminated in pressurized systems, which can improve the purity of CO_2 and reduce the cost of its capture [10,11].

Carbon dioxide utilization and storage require a gas stream with a concentration of CO_2 of 95% or more. The presence of impurities such as N_2, O_2, NO_x, SO_2, and other N- and S-containing gases affects the design, energy consumption, and costs of the CO_2 processing unit (CPU). Knowledge of pollutant formation and their concentration is, therefore, important for the design, performance, and cost of the CPU [12].

Several publications on the combustion of biomass in oxy-CFB systems are available in the open literature. A few of them provide some information emissions of NO_x, SO_2, and other pollutants. The main observations drawn from these studies are presented in Table 1.

In this study, combustion tests of three dedicated energy crops, namely miscanthus (*Miscanthus gigantheus*—a perennial gras with bamboo-like stems), Virginia mallow (*Sida hermaphrodita*—a perennial forb native to the eastern U.S.), and basket willow (*Salix viminalis*—a multi-stemmed shrub) were conducted in the bench-scale oxy-CFB reactor at 850 °C. On-line measurements of NO, NO_2, N_2O, NH_3, HCN, SO_2, and CO concentrations were taken during air- and oxy-fuel combustion. The influence of the chemical composition of oxidizing gas and biomass fuels on the emissions of above-mentioned pollutants was assessed. The results were compared to the emissions from wheat straw (agricultural biomass), Scots pine (woody biomass), and bituminous coal (reference fuel).

Table 1. Pollutant emissions from oxy-CFB combustion—a summary.

Reference	Test Facility and Operating Conditions	Fuels	Oxidizing Medium	Emissions Reported	Remarks
Tan et al. [13]	CFB, 800 kW_{th}, ~900 °C, limestone, Ca/S = 3	Wood pellets, different coals, fraction of biomass in fuel blends 20–50%wt.	O_2/CO_2 mixtures, 24–25% O_2, recycled flue gas	NO_x, SO_2, CO, CO_2, O_2	The addition of wood pellets did not have a notable impact on the combustion characteristics of tested fuels. Emissions of NO (per unit of energy (higher heating value) in the fuel) were in the range of 14–20 ng/J, and emissions of SO_2 varied from 35 to 95 ng/J. When the O_2 concentration in the flue gas exceeded 3.5%, emissions of CO dropped below 200 ppm.
Duan et al. [14]	CFB, 10 kW_{th}, 850 ± 10 °C	Rice husk, wood chips, dry wood flour, coal, biomass fraction in fuel 0–100%wt.	Air, 70% O_2/30% CO_2 mixture	NO, NO_2, CO, CO_2, O_2, SO_2	Emissions of NO were higher for biomass fuels than for coal. Oxy-combustion produced less NO than combustion in air. NO emissions in the air and oxy-fuel atmosphere increased with an increasing fraction of biomass in the fuel blend.
Wang et al. [15]	CFB, 10 kW_{th}, 800–900 °C	Corn straw, wheat straw, coal, 30% biomass in fuel blend	50% O_2/50% CO_2, 50% O_2/50% recycled flue gas	NO, N_2O, CO, CO_2, O_2, HCN	NO and N_2O emissions increased with an increasing excess O_2. An increase in the fraction of corn straw in the fuel blend caused an increase in emission factors of NO, N_2O, and HCN.
Varol et al. [16]	CFB, 850 and 915 °C, limestone, Ca/S = 2	Wood pellets, high-sulphur lignite, biomass/lignite ratio up to 60%	25, 30% O_2, CO_2	NO_x, SO_2, CO, CO_2	Increasing biomass share in the fuel blend had a negligible effect on NO_x emissions. Emissions of CO and SO_2 decreased with an increasing fraction of biomass in the fuel blend.
Sung et al. [17]	CFB, 30 kW_{th}, 750–840 °C	Sewage sludge, wood pellets	O_2 with recirculated flue gas	CO, NO	The lowest concentrations of CO (0.91%) and NO (14 ppm) were at 60% flue gas recirculation.
Nguyen et al. [18]	CFB, 100 kW_{th}, 845–905 °C, flue gas recirculation	Wood pellets, lignite, biomass fraction in fuel blend 50–100%wt.	21–29% O_2, CO_2	NO, SO_2, CO	An increase in biomass share caused a decrease in NO, SO_2 and CO concentrations. Oxy-combustion of pure biomass can produce negative CO_2 emissions of, approximately, −647 g/kW_{th}.
Moreno et al. [19]	CFB, 200 kW_{th}, 910 ± 10 °C, 835–852 °C, limestone	Coal, wheat straw, solid recovered fuel	O_2, recirculated flue gas	NO_x, SO_2, HCl	NO_x emissions of 359 mg/MJ_{th} for combustion of coal and 203 mg/MJ_{th} for co-combustion of coal with biomass. Emissions of SO_2 were 1.9–2.0 mg/MJ_{th}. Emissions of HCl were 1.8 mg/MJ_{th} for tests with no wheat straw and 3.9 mg/MJ_{th} for combustion of pure biomass.

While there are several publications on NOx emissions from the oxy-combustion of coal, relevant publications on the oxy-combustion of biomass are scarce [20]. There are no studies in the open literature on pollutant emissions during oxy-CFB combustion of energy crops such as *Miscanthus giganteus* and *Sida hermaphrodita*. This paper fills a gap in this field. This manuscript provides a comparative analysis of pollutant emissions during the combustion of energy crops and other fuels such as woody- and agro-biomass and bituminous coal. So, in this respect, it is similar to our previous work [12]. The test apparatus and methodology were the same to make sure the results can be compared.

2. Materials and Methods

2.1. Fuel Tested

Three different energy crops were selected for this study, *Miscanthus giganteus*, *Sida hermaphrodita*, and *Salix viminalis*, all originating from plantations in Poland. Agricultural biomass (wheat straw), woody biomass (Scots pine), and Polish bituminous coal were used as the reference fuels; the results of their combustion tests were already presented in reference [12]. Samples of tested fuels were milled and sieved to less than 0.2 mm for proximate and ultimate analyses. The basic characteristics of these fuels are summarized in Table 2. The ash yields for *Miscanthus giganteus* and *Sida hermaphrodita* are significantly higher than that for *Salix viminalis*. The energy crops tested in this study are characterized by a high content of volatile matter (VM). Their higher heating values (HHV) are in the range of 17.5 to 18.2 MJ/kg, typical for biomass fuels. Contents of sulphur and nitrogen in these crops are lower than in agricultural biomass (wheat straw) and coal. An additional advantage of these energy crops is a very low content of chlorine (less than 0.1%).

Table 2. Characteristics of tested fuels.

	Miscanthus giganteus	*Sida hermaphrodita*	*Salix viminalis*	Wheat Straw [12]	Scots Pine [12]	Bituminous Coal [12]
Proximate analysis (wt.%, db)						
Moisture	6.0	6.3	6.9	8.4	7.0	8.7
Ash	12.4	9.4	1.4	6.1	0.6	18.9
Volatile matter	75.8	76.7	76.3	68.3	76.8	26.8
Fixed carbon *	5.8	7.6	15.4	17.2	15.6	45.6
Higher heating value, MJ/kg	17.73	17.53	18.20	17.84	18.90	22.75
Ultimate analysis (wt.%, daf)						
Carbon	46.02	44.78	49.60	50.20	50.90	73.30
Hydrogen	5.38	5.19	6.00	5.80	5.70	4.30
Sulphur	0.03	0.02	0.03	0.08	0.01	2.30
Nitrogen	0.15	0.08	0.30	0.80	0.10	1.10
Chlorine	0.08	0.01	0.01	0.15	0.01	0.70
Oxygen *	48.34	49.92	44.06	42.97	43.28	18.30

Note: db—dry basis, daf—dry, ash free, * by difference to 100%.

2.2. Experimental Apparatus and Procedure

The experiments were conducted in the 12-kW bench-scale test apparatus under conditions similar to those experienced in a real CFB boiler at the Department of Thermal Machinery at the Czestochowa University of Technology. The apparatus consists of a combustion chamber (riser), a down-comer, a cyclone, and a loop-seal, and has been described in detail in references [21–25]. In this section, only the experimental conditions are briefly mentioned.

Mixtures of O_2 and CO_2 mixtures from gas cylinders were used to simulate the oxy-fuel combustion conditions. The concentrations of oxygen varied from 21% to 40%vol. All tests were carried out at a temperature of 850 °C without the addition of a sorbent for the capture of SO_2.

A 0.5 g sample of fuel was entered into the combustion chamber and burnt in the fluidized bed consisting of silica sand and ash particles. All fuel samples were prepared as pellets. The technique of pellets preparation is described in papers [23,24].

The composition of flue gases during the air and oxy-fuel combustion was measured by a portable FTIR spectrometer (Gasmet DX-4000) while O_2 was measured using a zirconium sensor analyzer (AMS Oxitrace). The volume fractions of NO, NO_2, N_2O, SO_2, CO, and other compounds were recorded every three seconds. The maximum error was less than 2%. Each experiment was repeated thrice minimum to guarantee a relative standard deviation below 5%.

In this study, temperature (850 °C), the superficial gas velocity (5 m/s), the mass of the fuel sample (0.5 g), and the mass of the inert material (0.3 kg) were constant. Only fuels and oxidizing atmospheres were different. Information collected during tests with a single fuel particle is sufficient to compare the influence of the combustion atmosphere on pollutant emissions. Our transient experiments with single biomass particles allowed us to collect data on instantaneous concentrations of emitted pollutants. These results are useful in assessing the behavior of biomass fuels in different combustion systems.

3. Results and Discussion

In this study, compositions of flue gases (NO_x, SO_2, and CO) during the air- and oxy-combustion of energy crops and reference biomass fuels and bituminous coal are reported and discussed.

3.1. Composition of Flue Gases during Air-Combustion

The effect of the fuel type on the instantaneous concentrations of NO, N_2O, SO_2, and CO during combustion in air is shown in Figure 1. The shapes of NO curves (Figure 1a) are similar for all energy crops, and their numerical values depend on the content of nitrogen and VM in the fuel. The highest emission of NO, approximately 60 ppm, was recorded for *Salix viminalis*. On the other hand, the lowest NO emission among all tested fuels was measured for *Sida hermaphrodita*, which is associated with the lowest nitrogen content in this fuel (see Table 2).

Emissions of N_2O (Figure 1b) were much lower than emissions of NO. Two peaks of N_2O concentration were observed during the combustion of biomass fuels. The first peak was seen during the combustion of VM and the second during the combustion of remaining char. Emissions of N_2O for energy crops did not exceed 8 ppm. The highest emissions of N_2O were recorded for wheat straw and the lowest for *Sida hermaphrodita*.

Nitrogen dioxide (NO_2) concentrations were below the detection limit for all tested fuels.

The concentration of SO_2 (Figure 1c) from the combustion of coal was much higher than that from the combustion of biomass. The highest value of the SO_2 instantaneous concentration was observed during the combustion of VM, and it was approximately 58 ppm. The highest SO_2 emissions for biomass fuels were detected for wheat straw. Concentrations of SO_2 for energy crops were below 4 ppm. The lowest SO_2 emission among all tested fuels was measured for *Sida hermaphrodita*, which is associated with the lowest sulphur content in this fuel (see Table 2).

Concentrations of CO in the flue gas during conventional combustion are shown in Figure 1d. The highest values of CO emissions were noticed during the char combustion, both for coal and biomass. The highest values of the instantaneous concentrations of CO were 27 ppm and 25 ppm for *Sida hermaphrodita* and *Miscanthus giganteus*, respectively. The release of CO lasted for 120–140 s for biomass, and 650 s for coal. Overall CO emissions were much higher for bituminous coal, which is associated with a higher content of carbon.

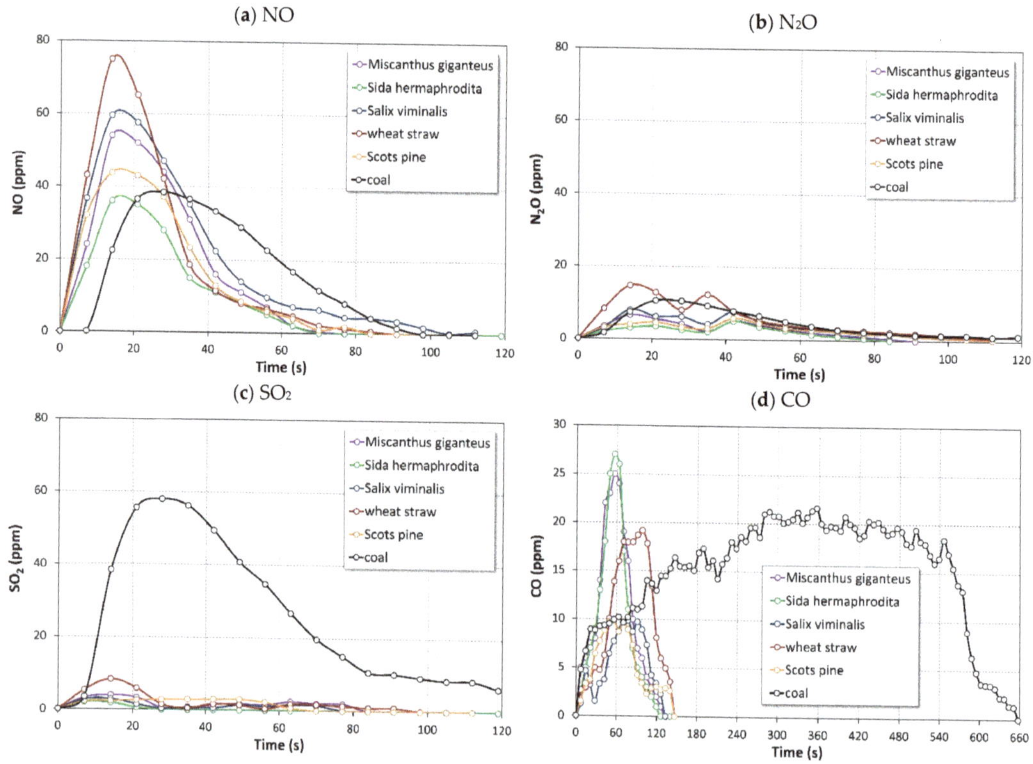

Figure 1. Effect of fuel type on instantaneous concentration of pollutants during air-combustion.

3.2. Composition of Flue Gases during Oxy-Combustion

The effect of the oxidizing atmosphere on the instantaneous concentrations of NO for all tested fuels is shown in Figure 2. The oxidizing atmosphere and type of fuel have a large influence on NO emissions. Concentrations of NO increased with increasing initial O_2 content in the O_2/CO_2 mixture. The highest values of NO concentrations were measured in the $40\%O_2/60\%CO_2$ atmosphere and the lowest in the $21\%O_2/79\%CO_2$ mixture. This is associated with the temperature of fuel particles [21]. A higher temperature of the fuel sample causes higher emissions of NO [12]. Additionally, a higher concentration of O_2 in the combustion chamber enhances the combustion of VM and leads to a rise in NO formation. This finding has been also confirmed by the results presented in previous papers [26,27].

The highest NO concentrations were observed in the combustion of VM, especially for biomass fuels. The highest values of the instantaneous NO emissions in the oxy-combustion of energy crops were detected for *Miscanthus giganteus* and the lowest for *Sida hermaphrodita*. During combustion of *Miscanthus giganteus* in the $40\%O_2/60\%CO_2$ atmosphere, the highest value of the NO concentration was 80 ppm, and it was approximately 30 ppm higher than that in the $21\%O_2/79\%CO_2$ atmosphere (Figure 2b). The maximum amount of NO for *Sida hermaphrodita* and wheat straw was approximately twice as high in the 40% O_2 mixture than that in the 21% O_2 mixture. The highest instantaneous concentrations of NO during air-combustion were similar to those in the $30\%O_2/70\%CO_2$ atmosphere for *Salix viminalis* and wheat straw. In other cases, combustion in the 30% O_2 mixture caused higher emissions of NO. The lowest emissions of NO during oxy-combustion were detected for the reference coal due to the lowest VM content in this fuel.

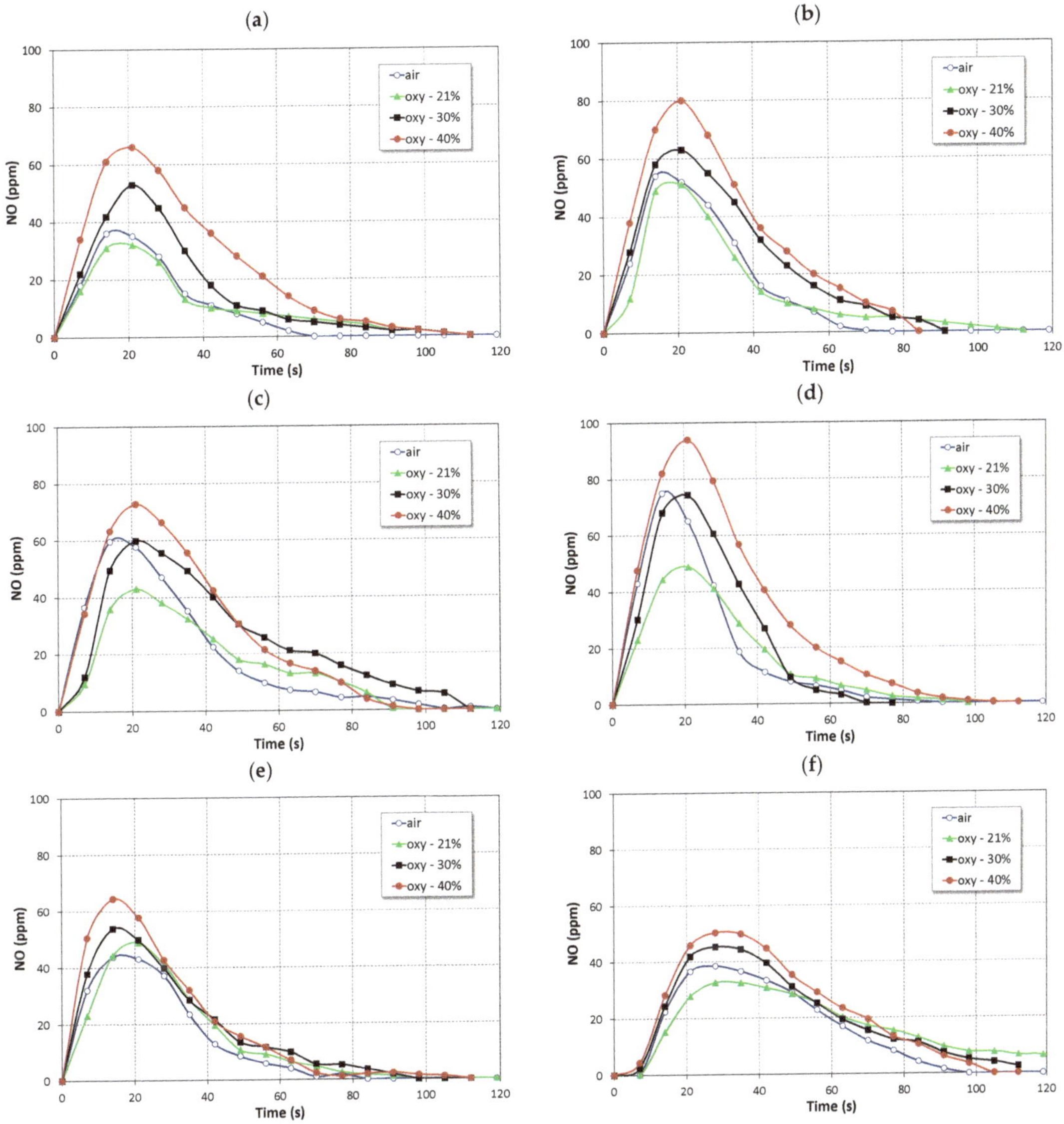

Figure 2. Effect of oxidizing atmosphere on instantaneous concentrations of NO for all tested fuels. (**a**) *Sida hemaphrodita*. (**b**) *Miscanthus giganteus*. (**c**) *Salix viminalis*. (**d**) Wheat straw [12]. (**e**) Scots pine [12]. (**f**) Bituminous coal [12].

Emissions of N_2O in the air- and oxy-combustion for all tested fuels are shown in Figure 3. Nitrous oxide concentrations were much lower in comparison with NO emissions. N_2O was formed at the same time as NO, which suggests that the formation of N_2O was proceeded by the direct oxidation of fuel-N instead of the reduction of NO [12]. The shapes of N_2O curves in air combustion are bimodal. The highest N_2O emissions were detected in the combustion of VM in the 21%O_2/79%CO_2 atmosphere. The highest value of N_2O concentration for energy crops, approximately 11 ppm, was observed for the combustion of *Miscanthus giganteus* (Figure 3b). As the oxygen concentration in the O_2/CO_2 atmosphere increased, N_2O concentrations decreased. The lowest N_2O emissions (below 8 ppm) for energy crops were observed for the combustion of *Sida hermaphrodita* both for air and oxy-combustion.

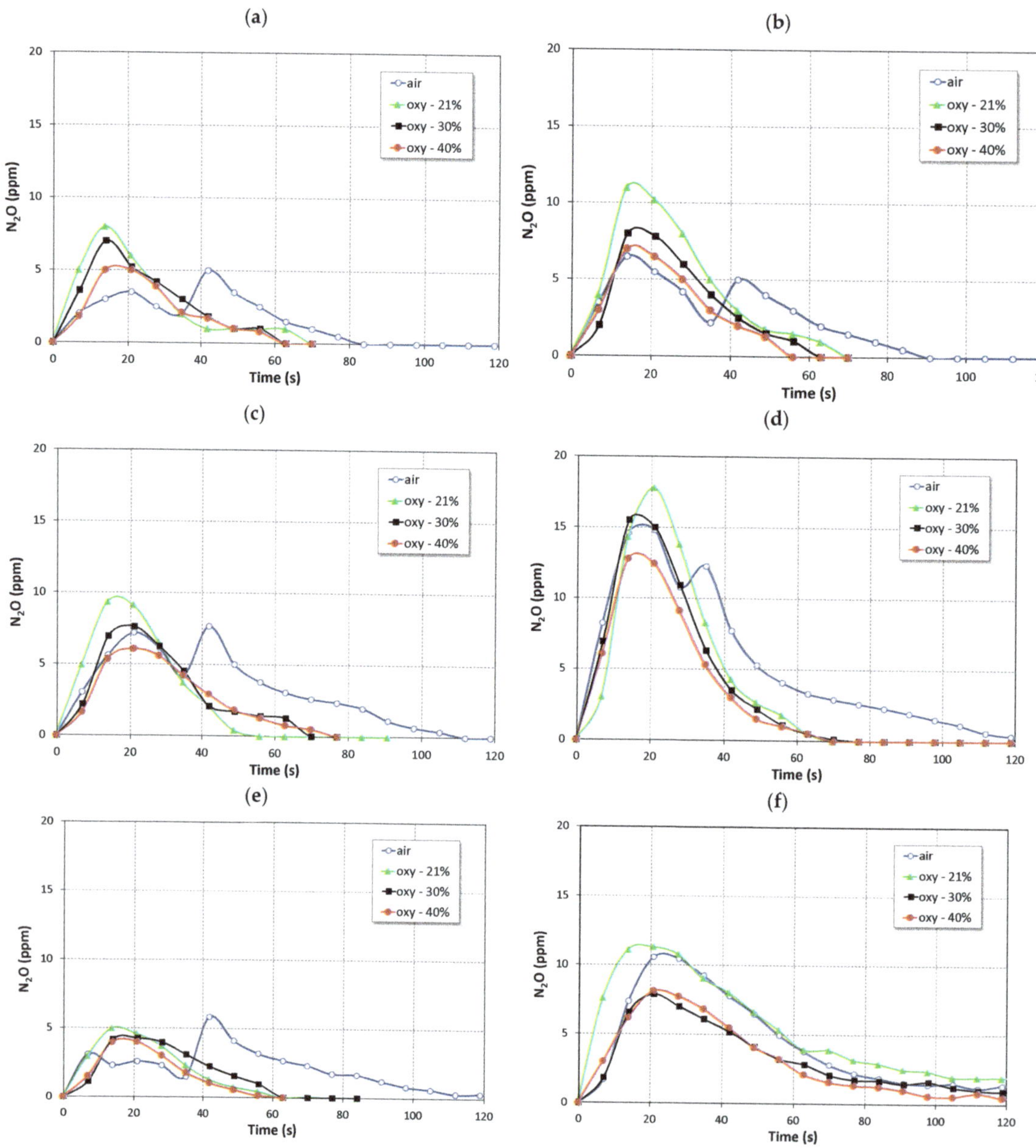

Figure 3. Effect of oxidizing atmosphere on instantaneous concentrations of N_2O for all tested fuels. (**a**) *Sida hemaphrodita*. (**b**) *Miscanthus giganteus*. (**c**) *Salix viminalis*. (**d**) Wheat straw [12]. (**e**) Scots pine [12]. (**f**) Bituminous coal [12].

HCN and NH_3 are the most important precursors of NO_x formation. Higher concentrations of HCN were generated in oxy-fuel combustion. The highest emissions of HCN were observed in 21%O_2/79%CO_2, mainly in the combustion of char. However, the highest values of the instantaneous HCN concentrations did not exceed 6 ppm for both energy crops and reference fuels. Analogous findings were reported in our previous study [23] and in reference [28]. The maximum ammonia emissions were observed during combustion in the mixture of 30%O_2 and 70%CO_2, but the values of the instantaneous concentrations were very low, below 3 ppm, for all samples. Large amounts of CO_2 inhibited the oxidation of NH_3 and HCN to NO_x [12].

Emissions of NO_2 were noted during the combustion of char only in the 21%O_2/79%CO_2 mixture. The maximum values of NO_2 concentration did not exceed 3 ppm for energy plants. This observation is consistent with the results obtained in references [25,28].

Figure 4 shows the influence of the oxidizing atmosphere on concentrations of SO_2 for energy crops and reference fuels.

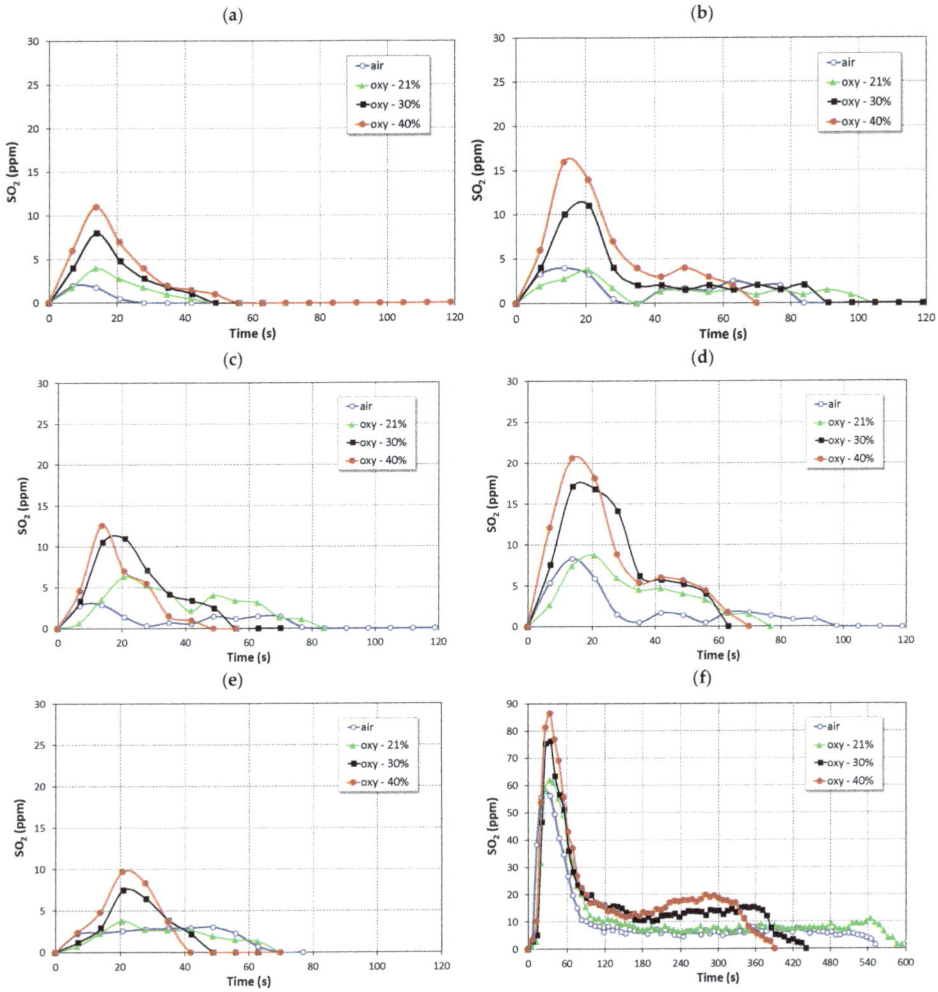

Figure 4. Effect of oxidizing atmosphere on instantaneous concentrations of SO_2 for all tested fuels. (**a**) *Sida hemaphrodita*. (**b**) *Miscanthus giganteus*. (**c**) *Salix viminalis*. (**d**) Wheat straw [12]. (**e**) Scots pine [12]. (**f**) Bituminous coal [12].

The amount of SO_2 formed during oxy-combustion of energy crops was much lower than that for the reference coal. For all tested samples, the maximum values of SO_2 were detected for the VM combustion period. An increase in SO_2 emissions was caused by an increase in the inlet concentration of O_2. The maximum value of SO_2 for energy crops was 16 ppm for *Miscanthus giganteus* in the oxy-40% atmosphere. However, the highest SO_2 concentration in the oxy-40% atmosphere was 68 ppm for coal. Oxy-combustion of *Miscanthus giganteus* in the mixture of 40% O_2 resulted in three times as high emissions of SO_2 than that in the mixture of 21% O_2. Comparable results for biomass and fossil fuels are reported in reference [28].

The effect of the oxidizing atmosphere on concentrations of CO during the combustion of energy crops and reference fuels is shown in Figure 5.

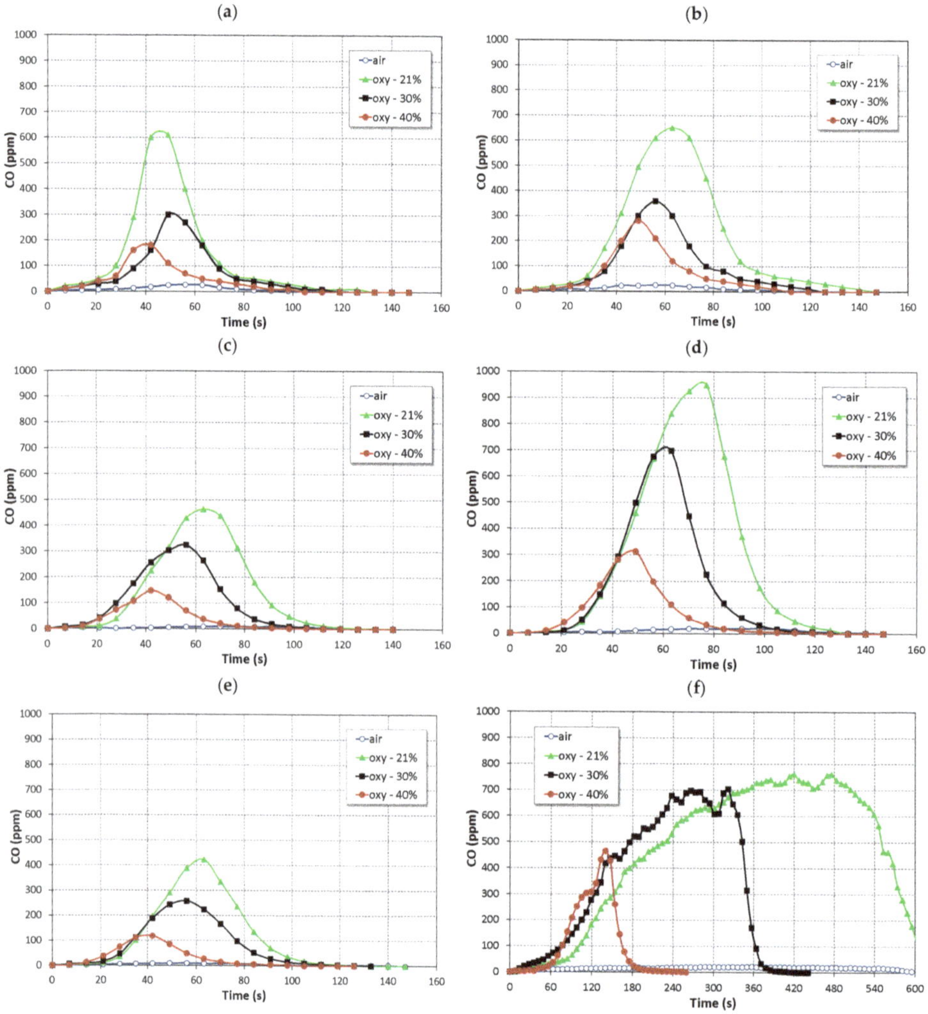

Figure 5. Effect of oxidizing atmosphere on instantaneous concentrations of CO for all tested fuels. (**a**) *Sida hemaphrodita.* (**b**) *Miscanthus giganteus.* (**c**) *Salix viminalis.* (**d**) Wheat straw [12]. (**e**) Scots pine [12]. (**f**) Bituminous coal [12].

The charts show the great influence of the oxidizing atmosphere on the amount of CO in exhaust gas. The oxy-combustion of energy crops in the mixture of 21% O_2 and 79 CO_2 caused a drastic increase in CO emissions compared to conventional combustion. The maximum values of the instantaneous CO concentrations in the oxy-21% atmosphere for energy crops were 650 ppm and 610 ppm, for *Miscanthus giganteus* and *Sida hermaphrodita*, respectively. This phenomenon could be described by the impact of the following reactions [12,28]:

$$CO_2 + C \leftrightarrow 2CO \tag{1}$$

$$CO_2 + H \leftrightarrow CO + OH \tag{2}$$

$$CO_2 \leftrightarrow CO + \frac{1}{2}O_2 \tag{3}$$

The first reaction, called Boudouard's reaction, appears to be predominant in the conversion of carbon monoxide under characteristic conditions for fluidized combustion [12,28]. The amount of CO in the flue gases decreased with an increase in the initial concentration of O_2. During combustion in the mixture of 40% O_2 and 60% CO_2, the maximum values of CO concentrations were thrice lower for *Sida hermaphrodita* and *Salix viminalis* than those in the oxy-21% mixture. Similar trends for biomass fuels were reported in references [25,28].

4. Conclusions

Three energy crops and reference fuels (agricultural and woody biomass and bituminous coal) were combusted in a bench-scale CFB, and the effect of the oxidizing atmosphere and fuel type on the composition of flue gases was studied. The most important conclusions from our study are as follows:

1. *Sida hermaphrodita* is the most environmentally friendly energy crop for both air and oxy-CFB combustion.
2. The instantaneous emissions of NO, N_2O, CO, and SO_2 for the combustion of energy crops in O_2/CO_2 environments are lower than those for combustion of agricultural biomass (wheat straw) and higher than those for combustion of woody biomass (Scots pine).
3. The instantaneous NO emissions during the combustion of energy crops in all atmospheres are higher than those of hard coal, which is due to the much higher content of volatile matter in renewable fuels.
4. The oxidation of volatile nitrogen compounds is behind high emissions of NO_x from energy crops burned in air and O_2/CO_2 mixtures.
5. The highest values of the instantaneous concentrations of NO and N_2O for energy crops during oxy-combustion are observed for *Miscanthus giganteus* and during air-combustion for *Salix viminalis*.
6. Combustion of energy crops in the oxy-21% atmosphere causes the lowest NO emissions and the highest N_2O emissions, which is associated with a lower temperature of burning fuel samples.
7. Emissions of CO for the combustion of energy crops in all atmospheres are much lower than those for the combustion of reference coal. It can be attributed to the higher content of carbon in fossil fuel.
8. Combustion of energy crops in the mixture of 21%O_2 and 79%CO_2 results in a very large increase in CO concentrations in the exhaust gas compared to conventional combustion.
9. As the initial concentration of oxygen in the O_2/CO_2 mixture increases, emissions of SO_2 and NO increase even though emissions of CO and N_2O decrease for energy crops and reference fuels.
10. Considering all emitted pollutants, the optimal atmosphere for the oxy-CFB combustion of energy crops should contain oxygen in the range of 21–30%vol.

Author Contributions: Conceptualization, M.K.-G., A.L. and T.C.; methodology, M.K.-G. and T.C.; formal analysis, M.K.-G.; investigation, M.K.-G. and T.C.; writing—original draft preparation, M.K.-G., A.L. and T.C.; writing—review and editing, M.K.-G., A.L. and T.C. All authors have read and agreed to the published version of the manuscript.

Funding: The work was funded by the statute subvention of Czestochowa University of Technology, Faculty of Mechanical Engineering and Computer Science (BS/PB-1-100-3011/2022/P).

Institutional Review Board Statement: Not applicable.

Informed Consent Statement: Not applicable.

Data Availability Statement: Not applicable.

Acknowledgments: We gratefully thank Krzysztof Wolski for his assistance in the experimental work.

Conflicts of Interest: The authors declare no conflict of interest.

References

1. Pradhan, P.; Mahajani, S.M.; Arora, A. Production and utilization of fuel pellets from biomass: A review. *Fuel Process. Technol.* **2018**, *181*, 215–232. [CrossRef]
2. Vassilev, S.V.; Baxter, D.; Andersen, L.K.; Vassileva, C.G. An overview of the chemical composition of biomass. *Fuel* **2010**, *89*, 913–933. [CrossRef]
3. Halford, N.G.; Karp, A. (Eds.) *Energy Crops. RSC Energy and Environment Series No. 3*; Royal Society of Chemistry: London, UK, 2011.
4. Anthony, E.J. Fluidized bed combustion of alternative solid fuels; status, successes and problems of the technology. *Prog. Energy Combust. Sci.* **1995**, *21*, 236–268. [CrossRef]
5. Giglio, R.; Castilla, N.J. The value proposition of circulating fluidized-bed technology for the utility power sector. *J. Southern African Inst. Mining Metall.* **2015**, *115*, 581–588. [CrossRef]
6. Basu, P. *Circulating Fluidized Bed Boilers. Design, Operation and Maintenance*; Springer: Cham, Switzerland; Heidelberg, Germany, 2015.
7. Natunen, M.; Jäntti, T.; Goral, D.; Nuortimo, K. First operating experiences of 55 MWe Konin and 205 MWe Połaniec CFB boilers firing 100% biomass. In Proceedings of the Paper presented at PowerGen Europe conference, Vienna, Austria, 4–6 June 2013.
8. Giglio, R.; Jäntti, T. Tees Renewable Energy Plant. *Mod. Power Syst.* **2017**, 21–26.
9. Jia, L.; Tan, Y.; Anthony, E.J. Emissions of SO_2 and NOx during oxy-fuel CFB combustion in a mini-circulating fluidized bed combustion reactor. *Energy Fuels* **2010**, *24*, 910–915. [CrossRef]
10. Li, L.; Duan, L.; Yang, Z.; Zhao, C. Pressurized oxy-fuel combustion characteristics of single coal particle in a visualized fluidized bed combustor. *Comb. Flame* **2020**, *211*, 218–228. [CrossRef]
11. Liang, X.; Wang, Q.; Luo, Z.; Eddings, E.; Ring, T.; Li, S.; Yu, P.; Yan, J.; Yang, X.; Jia, X. Experimental and numerical investigation on nitrogen transformation in pressurized oxy-fuel combustion of pulverized coal. *J. Clean. Prod.* **2021**, *278*, 123240. [CrossRef]
12. Kosowska-Golachowska, M.; Luckos, A.; Kijo-Kleczkowska, A. Pollutant emissions during oxy-fuel combustion of biomass in a bench scale CFB combustor. *Energies* **2022**, *15*, 706. [CrossRef]
13. Tan, T.; Jia, L.; Wu, Y. Some combustion characteristics of biomass and coal cofiring under oxy-fuel conditions in a pilot-scale circulating fluidized combustor. *Energy Fuels* **2013**, *27*, 7000–7007. [CrossRef]
14. Duan, L.; Duan, Y.; Zhao, C.; Anthony, E.J. NO emission during co-firing coal and biomass in an oxy-fuel circulating fluidized bed combustor. *Fuel* **2015**, *150*, 8–13. [CrossRef]
15. Wang, X.; Ren, Q.; Li, W.; Li, H.; Li, S.; Lu, Q. Nitrogenous gas emissions from coal/biomass co-combustion under a high oxygen concentration in a circulating fluidized bed. *Energy Fuels* **2017**, *31*, 3234–3242. [CrossRef]
16. Varol, M.; Symonds, R.; Anthony, E.J.; Lu, D.; Jia, L.; Tan, Y. Emissions from co-firing lignite and biomass in an oxy-fired CFBC. *Fuel Process. Technol.* **2018**, *173*, 126–133. [CrossRef]
17. Sun, J.-H.; Back, S.-K.; Jeong, B.-M.; Kim, J.-H.; Choi, H.S.; Jang, H.-N.; Seo, J.-C. Oxy-fuel co-combustion of sewage sludge and wood pellets with flue gas recirculation in a circulating fluidized bed. *Fuel Process. Technol.* **2018**, *172*, 79–85.
18. Nguyen, H.K.; Moon, J.-H.; Jo, S.-H.; Park, S.J.; Seo, M.W.; Ra, H.W.; Yoon, S.-J.; Yoon, S.-M.; Song, B.; Lee, U.; et al. Oxy-combustion characteristics as a function of oxygen concentration and biomass co-firing ratio in a 0.1 MWth circulating fluidized bed combustion test-rig. *Energy* **2020**, *196*, 117020. [CrossRef]
19. Moreno, J.; Hornberger, M.; Schmid, M.; Scheffknecht, G. Oxy-fuel combustion of hard coal, wheat straw, and solid recovered fuel in a 200 kW_{th} calcium looping CFB calciner. *Energies* **2021**, *14*, 2162. [CrossRef]
20. Meng, X.; Rokni, E.; Zhou, W.; Qi, H.; Sun, R.; Levendis, Y.A. Emissions from oxy-combustion of raw and torrefied biomass. *J. Energy Resour. Technol.* **2020**, *142*, 122307. [CrossRef]
21. Kosowska-Golachowska, M.; Kijo-Kleczkowska, A.; Luckos, A.; Wolski, K.; Musiał, T. Oxy-combustion of biomass in a circulating fluidized bed. *Arch. Thermodyn.* **2016**, *37*, 17–30. [CrossRef]

22. Kosowska-Golachowska, M.; Luckos, A.; Musiał, T.; Kijo-Kleczkowska, A.; Wolski, K.; Środa, K. NO_x and N_2O emissions during oxy-fuel combustion of bituminous coal and lignite in a circulating fluidized bed combustor. In Proceedings of the 13th International Conference on Fluidized Bed Technology, Vancouver, BC, Canada, 10–14 May 2021; pp. 587–592.
23. Kosowska-Golachowska, M.; Luckos, A.; Kijo-Kleczkowska, A.; Musiał, T.; Wolski, K.; Środa, K. Gaseous emissions during oxy-fuel combustion of sewage sludge in a circulating fluidized bed. *Powder Technol.* **2020**, *371*, 209–216. [CrossRef]
24. Kijo-Kleczkowska, A.; Środa, K.; Kosowska-Golachowska, M.; Musiał, T.; Wolski, K. Combustion of pelleted sewage sludge with reference to coal and biomass. *Fuel* **2016**, *170*, 141–160. [CrossRef]
25. Wolski, K. Experimental investigation into oxy-fuel combustion of agro and wooden biomass in a circulating fluidized bed. Ph.D. Thesis, Czestochowa University of Technology, Czestochowa, Poland, 2016. (In Polish).
26. Czakiert, T.; Sztekler, K.; Karski, S.; Markiewicz, D.; Nowak, W. Oxy-fuel circulating fluidized bed combustion in a small pilot-scale test rig. *Fuel Processing Technol.* **2010**, *91*, 1617–1623. [CrossRef]
27. Jankowska, S.; Czakiert, T.; Krawczyk, G.; Borecki, P.; Jesionowski, Ł.; Nowak, W. The effect of oxygen staging on nitrogen conversion in oxy-fuel CFB environment. *Chem. Process Eng.* **2014**, *35*, 489–496. [CrossRef]
28. Lasek, J.A.; Janusz, M.; Zuwała, J.; Głód, K.; Iluk, A. Oxy-fuel combustion of selected solid fuels under atmospheric and elevated pressures. *Energy* **2013**, *62*, 105–112. [CrossRef]

energies

MDPI

Article

Consequence of Blowby Flow and Idling Time on Oil Consumption and Particulate Emissions in Gasoline Engine

Vincent Berthome *, David Chalet and Jean-François Hetet

Ecole Centrale de Nantes, LHEEA Laboratory (ECN/CNRS), 44321 Nantes, France
* Correspondence: vincent.berthome@ec-nantes.fr

Abstract: Pollutant emission standards and, in particular, those concerning particles from an internal combustion engine (ICE) are becoming increasingly restrictive. Thus, it is important to determine the main factors related to the production of particulate matter. In this article, the phenomenon of oil sweeping by the blowby gases between the rings/piston/cylinder is investigated. First, a blowby gas simulation model based on experimental results from a Turbocharged Gasoline Direct Injection (TGDI) is developed. From this model, it is possible to characterise the amount of oil swept by the blowby gases. This depends on the endgap position of both the compression and sealing rings. It also depends on the intensity of the blowby flow rate, which is highest at low rpm and high load. At 1500 rpm and full load, this flowrate exceeds 25 mg.cycle^{-1}. From this result, it is possible to quantify the amount of oil swept by these gases as a function of the endgap position. For $\theta_{rings} = 180°$, the quantity of oil swept rises to 20 µg.cycle^{-1} while for $\theta_{rings} = 30°$, this decreases to 6 µg.cycle^{-1}. The oil concentration of the blowby gas has a direct impact on the particulate emissions because the oil concentration of the backflow gas is inversely proportional to the blowby gas flowrate. As the backflow gases return to the cylinder, the oil oxidises and produces particles. Therefore, it is essential to control the oil concentration of the backflow gases. In addition, the simulation model shows the blowby flowrate becomes negative and decreases to −3.4 mg. cycle^{-1} in idle conditions. The amount of oil swept by the blowby is no longer directed towards the oil pan, but towards the piston crown. This phenomenon of oil storage of the piston crown in idle condition is proportional to the duration of the idle time. In order to confirm these results, experimental tests are carried out on a TGDI engine. It appears that when the idling time changes from 0 s to 7 s between two strictly identical accelerations, the level of particulate emissions is multiplied by 1.3. When the idling time changes from 0 s to 22 s between two strictly identical accelerations, the level of particulate emissions is multiplied by 3. These results confirm the mechanism of oil storage at idle highlighted by the simulation model.

Keywords: particle; oil; blowby; backflow; endgap; ring; gasoline; idle

Citation: Berthome, V.; Chalet, D.; Hetet, J.-F. Consequence of Blowby Flow and Idling Time on Oil Consumption and Particulate Emissions in Gasoline Engine. *Energies* **2022**, *15*, 8772. https://doi.org/10.3390/en15228772

Academic Editors: Tomasz Czakiert and Monika Kosowska-Golachowska

Received: 13 October 2022
Accepted: 18 November 2022
Published: 21 November 2022

1. Introduction

Human activities have an impact on the environment by producing large quantities of greenhouse gases and polluting emissions. In the field of transport, the main pollutants linked to internal combustion engines are CO, NOx, unburnt hydrocarbons, and particles. The size of these particles can be less than a few tens of nanometers and therefore they can penetrate the bronchial tubes, reach the pulmonary alveoli, and have adverse effects on human health. It is therefore necessary to reduce particulate emissions from an internal combustion engine. In Europe, the Euro 6d regulation limits these emissions to 6.10^{11} particles by km and 4.5 mg of particles by km with the Worldwide Harmonized Light Vehicle Test Procedure cycle (WLTP) [1], for particles larger than 23 nm. The particle formation is a complex mechanism. It consists of a solid part composed of carbon and hydrogen containing a soluble organic fraction such as unburnt hydrocarbons, oxygenated derivatives or polyclinic aromatic hydrocarbons (PAH), or containing non-organic elements, such as mineral derivatives or metallic residues [2]. A particle arises from the collision between two PAH of pyrene ($C_{16}H_{10}$). This is called nucleation [3–8]. For internal combustion engines, the main factor generating particulate matter is the high-temperature combustion or pyrolysis of unburnt fuel [9]. Particle fluctuation at steady-state regimes was observed by Thawko, et al., [10] on a gasoline engine and by Swanson, et al., [11] on a diesel engine. Berthome, et al., [12,13] found variations in particulate emissions from a TGDI engine up to 80% on strictly identical transients. Amirante, et al., [14] found that the predicted soot particles mass was lower than experimental values and they suggested that other sources than fuel combustion should be considered, such as oil consumption. The main source of particulate emissions after unburnt fuel is oil consumption [15–18]. This consumption has several origins, such as the expulsion of oil through the piston rings into the combustion chamber, the evaporation of the oil film, and the transport of oil through the valve guides or via the blowby gases [19–22]. During a transient, Berthome, et al., [12,20] determined that 57% of the particulate emissions from an ICE were related to the variation of richness, while 31% were related to blowby, and 12% to backflow.

Blowby gases formation:

Blowby and backflow gases therefore have a very significant impact on particulate emissions. For example, an TGDI without blowby gases reintroduced at the intake emits 1.5 times less particle emissions than the same engine in normal configuration [20]. These gases, resulting from combustion, pass through the labyrinths of the cylinder/piston, ring, and grooves to the oil pan. These hot gases absorb oil droplets as they pass through the cylinder. Most engines use an oil sump. This means that the oil falls back into the crankcase by gravity after lubricating the components such as the crankshaft bearings, camshafts, valves, turbocharger, etc. A pump immersed in this housing sucks the oil and delivers it to the components to be lubricated after passing through the oil filter. With this oil falling by gravity, as well as the spray lubrication of the cylinder walls and moving parts, an oil mist is created which is partially absorbed by the hot gases of the blowby. If nothing is done, the pressure in the lower crankcase increases and can lead to problems with sealing, lubrication, etc. It is therefore necessary to evacuate these gases for the proper functioning of the engine. Since the introduction of anti-pollution standards, these gases are no longer released into the atmosphere but are reinjected at the intake. The oil in these gases must be removed before re-introducing them into the intake manifold [23].

The blowby oil sweep phenomenon:

Min, et al., [24] found that the blowby gases had the particularity of sweeping the oil stored between the compression and sealing rings of a piston. The position of the first two rings, characterized by the position of the "endgap", influences the flow rate and oil concentration of the blowby gases [24–27]. Thirouard, et al., [28] deduce that the amount of oil returned to the oil pan (Q_{oil}) depends on the blowby flow rate (Q_{blowby}) and the position of the "endgap" of these rings (θ_{rings}), cf., Equations (1) and (2).

$$Q_{oil}=Q_{blowby}\left(\frac{1}{2\pi}\right)\frac{\mu_{air}}{\mu_{oil}}\cdot\frac{3\,K_1}{K_2}\left((2\pi-\theta_{rings})\left(1-e^{\frac{-K_2\theta_{rings}}{C_1(2\pi-\theta_{rings})}}\right)+\theta_{rings}\left(1-e^{\frac{-K_2(2\pi-\theta_{rings})}{C_1\theta_{rings}}}\right)\right), \tag{1}$$

with

$$C_1 = Q_{blowby}\frac{\mu_{air}}{\mu_{oil}}\cdot\frac{3}{2\pi(h_{air})^2R_{land}} \tag{2}$$

where R_{land} is the clearance between piston radius and cylinder radius, K_1 and K_2 constants are mainly related to the engine speed, and h_{air} is the clearance between the piston and the cylinder.

A ring is a split elastic ring made mainly of steel. This endgap is necessary to counteract thermal deformations that cause a ring dilatation. Depending on the chemical properties of the ring, this cutting clearance is more or less important. When the endgaps of the compression and sealing rings are opposite each other at 180°, the blowby gases sweep over the entire inter-ring area, see Figure 1a. This means that a maximum amount of oil stored in this zone is redirected to the oil pan. This phenomenon is called "blow-down effect".

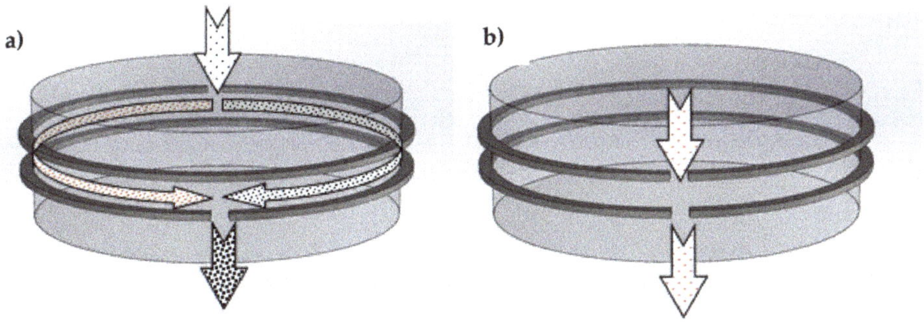

Figure 1. Oil removed by the blowby gas vs. endgap position of rings.

The gases, reintroduced into the cylinder during the next cycle, and which will pass through this zone again in the other way via the backflow phenomenon, will be lightly charged with oil and, consequently, the associated level of particles will be limited. On the other hand, if the endgap positions are close, for example, 0°, the area swept by the gases will be very limited and there will be a lot of oil left in this area, see Figure 1b. The backflow gases will then recharge with oil and redirect it to the cylinder, resulting in a higher particle level in the next cycle. Nakashima, et al., [29] and Agarwal, et al., [30] measured very significant variations in oil flow as a function of endgap position, up to 400% depending on the piston design.

Rings dynamic behaviour:

The ring dynamic behaviour is a complex mechanism. A piston ring rises and falls in its groove several times per cycle. This is due to the different forces it undergoes. Tian, et al., [31] determined that the position of the ring depended on the gas pressure upstream and downstream of the ring, inertial forces, frictional forces, the pressure distribution on the side of the ring, and the contact surface of the ring on the cylinder. However, under special engine conditions, that is to say under heavy load and for high rotational speed, the first two rings can sometimes move radially inwards of the piston grooves and allow unburned gases to pass directly into the oil pan via the scraper ring [32]. Tian [33] and Chen [34] showed that this dynamic behaviour, called radial collapse, is very sensitive to the pressure above the ring and to the tilt angle of the ring. When the ring collapses, it generates more blow-by. Tian [31] demonstrated that it can be limited via the shape of the rings. Rabuté and Tian [35] found out that blow-by is sensitive to the choice of ring materials. Wroblewski and Koszalka [36] measured the influence of various anti-wear coatings on frictional losses on the rings. Wroblewski and Iskra [37] demonstrated that the asymmetrical shape of the rings impacts the amount of oil scraped into the combustion chamber during the compression and exhaust stroke. Zarenbin, et al., [38] studied the impact of piston ring mobility on the blow-by gas and determined that the movements of the rings in the grooves noticeably affect the gas escape into the crankcase. Turnbull, et al., [39] showed that the power losses due to gas leakage can be more important than frictional losses.

There are various singularities in the dynamics of the rings, such as axial flutter and radial flutter, but the one that most impacts the blowby phenomenon is rotational movement. The rotation of the rings is due to two phenomena: micro-scratches in the cylinder, and piston oscillations. In fact, the rings do not move back and forth in a radial direction, but by reversing them in the cylinder bore, the piston changes its support from one cylinder wall to the other if it is on the thrust or anti-thrust side. This occurs both at Top Dead Center (TDC) of the piston and at Bottom Dead Center (BDC). This results in a radial displacement of the ring in its groove. This leads to the rotation of the ring in relation to the cross pass and the honing structure. Schneider, et al., [40] and Min, et al., [24] measured ring rotation of up to 10 rpm. This depends on the load and engine speeds [41]. Thirouard, et al., [26] studied the impact of the position of the rings and the amount of oil in the space between the first two rings (Land 2). It appears that when the rings are free to rotate, the amount of oil is very unstable, whereas it remains constant when they are locked in rotation.

Blowby gas simulation:

In order to better characterise the quantity of oil swept by the blowby gases and its impact on particle emissions, it appears necessary to create a simulation model of these gases. The novelty of this article is that it is possible to quantify the amount of oil swept by these gases from a simulation model and as a function of the position of the piston rings and the engine operating points. The first part of this paper is devoted to the calibration and analysis of the simulation model based on experimental tests, while the second part refers to the analysis of a particular engine operating point that is idling.

2. Materials and Method

2.1. Description of Engine Bench

The tests were carried out on a three-cylinder spark-ignition engine. It is a turbocharged direct-injection engine. Its characteristics are specified in Table 1. It was braked by a dynamic HORIBA HT 250 bench managed by a SPARC control unit. The engine was instrumented with type K temperature sensors, 0–3 bar static pressure sensors, two Kistler type 4049B dynamic pressure sensors for the intake and the exhaust (frequency = 60 kHz), one HBM type 40 torquemeter (frequency = 10 kHz), three AVL type ZI33 in-cylinder pressure sensors (frequency = 150 kHz), and one ETAS ES430 air-fuel ratio sensor (frequency = 2 kHz).

Table 1. Engine main features.

Engine Main Features		Main Parameters of Piston Rings	
Engine	4-stroke, TGDI, 3-cylinder	Piston diameter	74.45 mm
Bore × Stroke	75 mm × 90.48 mm	Ring 1 axial height	1.2 mm
Displacement	1199.9 cm^3	Ring 2 axial height	1 mm
Valves	12	Ring 3 axial height	2 mm
Compression ratio	10.5: 1	Ring 1 end gap	0.2 mm
Maximum Power	96 kW @ 5500 rpm	Ring 2 end gap	0.4 mm
Maximum Torque	230 Nm @ 1750 rpm	Ring 3 end gap	0.2 mm

The INCA Software (version 7.1.10/3) controlled the engine ECU.

All the sensors were connected to a National Instrument fast acquisition box. The STARS software from HORIBA made it possible to automatically control the dynamic bench, and it could create different cycles. The PR-L804 fan from Dynair cooled the engine radiator, and the Fumex FB110 fan cooled the intake heat exchanger. Both fans were controlled by STARS software, which modified their rotation speeds according to the speed of the vehicle during a cycle. This equipment created test conditions close to the real ones. The pollutant emissions CO and HC were measured with a 3200 CAPELEC device, the NOx emission with an ECM NOxCANt sensor, and particle emission with a Pegasor Particle Sensor (PPS). The setup is shown in Figure 2.

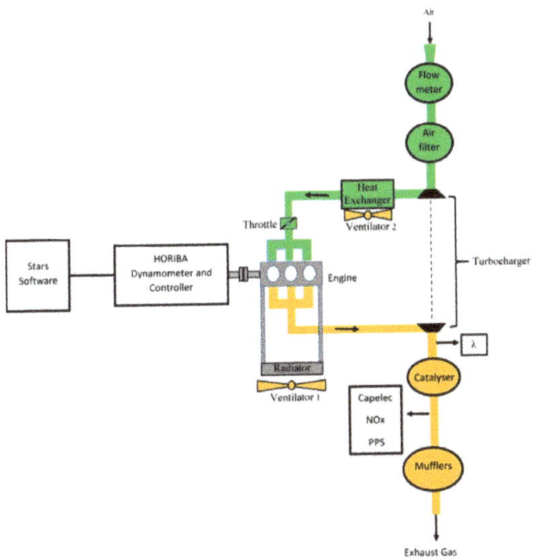

Figure 2. Experimental setup.

3. Blowby Simulation in Steady State

3.1. Building a Blowby Simulation Model

As a first step, it is necessary to model the blowby gases of the engine described in Table 1. In this way, the Gamma Technology simulation software, GT-suite 2021, which has a "RingPack" module (specially created to model piston/cylinder/ring friction and to model the pressures and flows of gases circulating through the rings and piston grooves) was used. Only a single cylinder simulation model was created. Indeed, the objective was to create a simulation model of the phenomenon of oil scavenging by the blow-by gases according to the value of the final gap of the first two rings of one of the cylinders of the engine used. This single cylinder has all the technical characteristics of the engine used, such as the cylinder and piston diameters, the length of the connecting rod, the depth, height, and location of the piston grooves and the rings. Some of these characteristics are shown in Table 1. The input parameters required for this model were the temperature and pressure inside the cylinder during a cycle, see Figure 3. The latter was derived from experimental tests, while the temperature was approximated by the extensive literature on the subject. Indeed, temperature is a key factor, as it determines the proportion of an object that expands according to its physical properties and, in particular, its coefficient of expansion. Piston, ring and cylinder wall temperatures are proportional to the specific power developed by the engine and therefore depend on the engine load and speed [42]. On a single-cylinder diesel engine at 25% load and a speed of 3600 rpm, Abril, et al., [43] measured a compression ring temperature close to 200 °C during the combustion and expansion phases. On a single-cylinder, spark-ignition engine at 75% load and 3500 rpm, Thiel, et al., [42] determined that the temperature of the compression ring could reach 210 °C during these same phases. Husberg, et al., [44] found piston crown temperature variations of up to 75 °C between 25% and 50% load on a single-cylinder diesel engine. Taking all these elements into account, a coefficient of thermal expansion of the compression ring equal to 1×10^{-5} K^{-1} with a maximum temperature of 250 °C at 6000 rpm in full load is assumed. The temperature of the compression ring changes linearly with the engine speed.

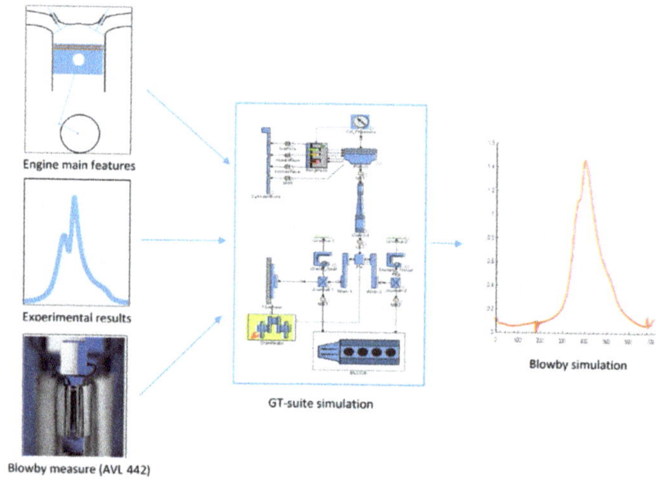

Figure 3. Simulation model development with GT-suite.

Model calibration:

The model calibration was based on experimental tests perform by a car manufacturer development department. These tests were carried out under full load and steady state on an engine strictly identical to the one used, see Table 1. The blowby rate was measured with an AVL 442 blowby meter. This device determines a flowrate from the measurement of a pressure difference generated by the blowby gases through a change in cross-section in the blowby gas intake duct. The results are displayed by the device in l.min^{-1} for a pressure of 1000 mbar and a temperature of 25 °C [45], see Table 2.

Table 2. Blowby flow measurement for the engine at full load.

Engine speed (rpm)	1000	1500	2000	2500	3000	3500	4000	4500	5000	5500	6000
Qv (l.min^{-1})	31	52	51	52	51	49	49	49	51	49	47
Qm (mg.cycle^{-1})	24.1	26.9	19.9	16.2	13.2	10.9	9.51	8.46	7.92	6.92	6.08

Study of the position of the compression ring under full load

Considering only the axial motion of the ring and applying the fundamental principle of dynamics, the equation is:

$$m\frac{d2h}{dt2} = F_p+F_i+F_{fr}+F_{oil}+F_a \tag{3}$$

where $\boldsymbol{m}$ is the ring mass. The oil film pressure forces (F_{oil}) and adhesion forces (F_a) are not significant compared to the intensity of the forces pressure forces (F_p), inertial forces (F_i), and friction forces (F_{fr}) [46]. The position of the ring in the piston groove (h) is therefore summarised by the variation of the forces shown in Figure 4. At low rotational speeds, i.e., between 1500 rpm and 3500 rpm, the pressure forces, F_p, are greater than the inertial forces, F_i, and therefore the compression ring remains in a low position in the piston groove throughout the cycle. From 4000 rpm, the inertial forces take advantage over the pressure forces F_p, which makes the ring move in its groove.

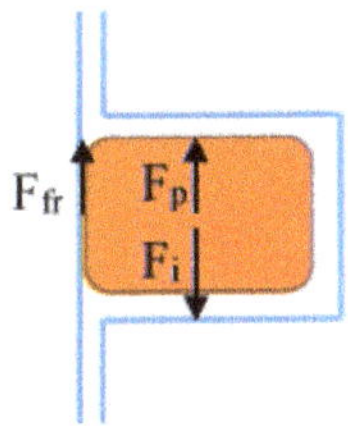

Figure 4. Set of Forces supported by the ring.

When the cylinder pressure is lower than the crankcase pressure, the ring may move in its groove. This is the case at 1000 rpm, since the pressure in the cylinder is lower than the atmospheric pressure when the exhaust gases are removed.

Study of the position of the sealing ring under full load

The same analysis can be made on this ring and, given the lower pressures involved, it moves axially at each cycle and whatever the rotation speed.

Simplification of the calibration

Usually, the endgap value of the sealing ring is much higher than that of the compression ring. In this case, it is twice as large (0.4 mm versus 0.2 mm). As a result, the blowby gas flow is mainly related to the position of the compression ring (Ring 1). Between 1500 rpm and 3500 rpm, the compression ring stays fixed on its base. In this speed range, the only possible path for the blowby gases is through the endgap of the compression ring. Therefore, the calibration was focused on this speed range and then extended to the entire engine speed range.

Analysis at 2000 rpm and full load

Figure 5 shows the simulation results at 2000 rpm and full load. The pressures between the rings are visible in the upper and middle graphs. The pressure above the compression ring (Land 1) is identical to the pressure in the cylinder. Its maximum is 77 bars at 400° CA.

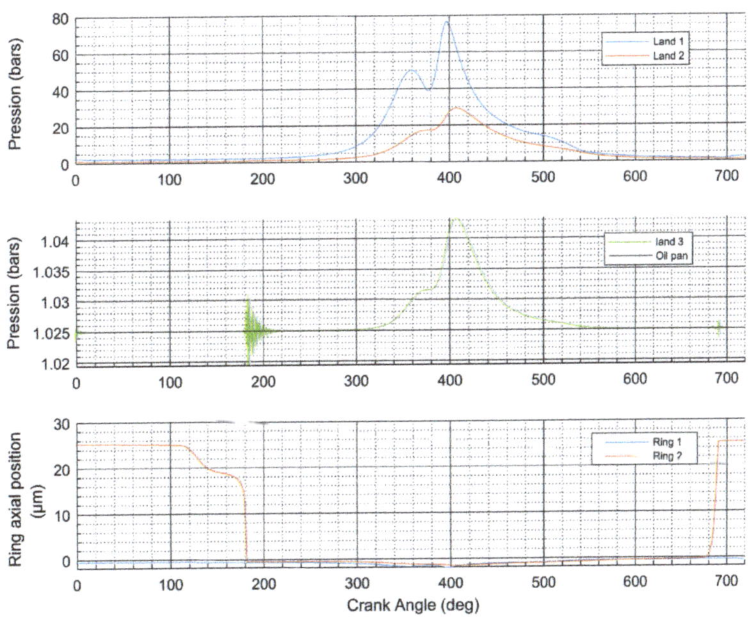

Figure 5. Evolution of the pressures and axial positions of the piston rings at 2000 rpm and full load.

The maximum pressure between the compression and sealing rings (Land 2) is 29 bars. The pressure between the oil seal and scraper rings (Land 3) varies between 1025 mbar and 1045 mbar, while the pressure in the crankcase was stabilised at 1025 mbar. From this data and the inertial effects due to the speed of the piston displacement, it is possible to characterise the axial evolution of the compression and sealing rings, which is visible on the lower graph of Figure 5. As expected, the compression ring stays fixed on the lower part of the groove during the whole cycle, while the sealing ring is in the upper position of its groove during the end of the exhaust gas expulsion stroke and during the whole intake stroke. Outside this range, it stays in contact with the lower part of its groove. The evolution of the blowby flowrate can be seen in Figure 6. This one becomes significant from the end of the compression stroke to the end of the expansion stroke. A maximum was reached as 40° after TDC and amounted to 1.45 g.s^{-1}. Finally, the cumulative blowby flowrate for one cycle reached 20.26 mg.cycle^{-1}. This result is very close to the experimental tests where the cumulative blowby flowrate for this same operating point was measured at 19.90 mg.cycle^{-1} (cf., Table 2). This simulation model allows us to better understand the ring dynamics and to estimate the evolution of the pressures that prevail in this ring/piston/cylinder area. Above all, thanks to the characterisation of the blowby flowrate over a complete cycle, it is possible to determine which engine strokes are preponderant in the phenomenon of oil sweeping towards the oil pan.

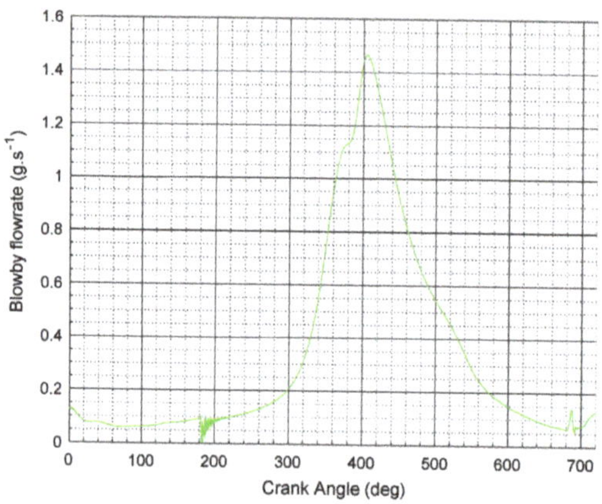

Figure 6. Evolution of blowby flowrate at 2000 rpm and full load.

Simulation over the entire engine load

From this model, and by adding different engine operating points performed on the engine bench, it is possible to draw a complete map of the blowby flowrate as a function of engine speed and engine load.

The result is shown in Figure 7. It is very clear that the flowrate is higher than 18 mg.cycle^{-1} in a restricted area, i.e., at low speed and high load. Conversely, outside this zone, the flow rate remains below 12 mg.cycle^{-1}.

3.2. *Analysis of Oil Flow Carried by Blowby Gases*

From these results and Equation (1), it is possible to determine the amount of oil swept by the blowby gases as a function of the position of the rings. To do this, it is necessary to estimate the amount of oil present in the Land 2. This quantity of oil depends on two factors: the first is the thickness of the oil film deposited by the ring, which varies by a few micrometers [47]. The second is the oil that is pumped out of the piston grooves and back into this area. This quantity of oil depends on the engine load and inertial effects, e.g., the rotation speed. As an example, Thirouard [48] measured an oil film height of up to 40 µm at 3500 rpm and half load on a single cylinder spark ignition engine and 20 µm at 2800 rpm and low load on a single cylinder diesel engine. In order to quantify the results, it is necessary to suppose an oil film height. The characteristics of the engine used are close to a diesel engine, therefore the assumption of an oil film of 20 μm will be retained. Figure 8 shows the amount of oil swept by the blowby gases at 1500 rpm for two endgap positions with an assumed oil film height of 20 μm. When the endgaps of the first two rings are opposite, i.e., at 180° (cf Figure 1a), the amount of oil swept by the blowby gases is maximum and can reach 20 μg.cycle^{-1}.

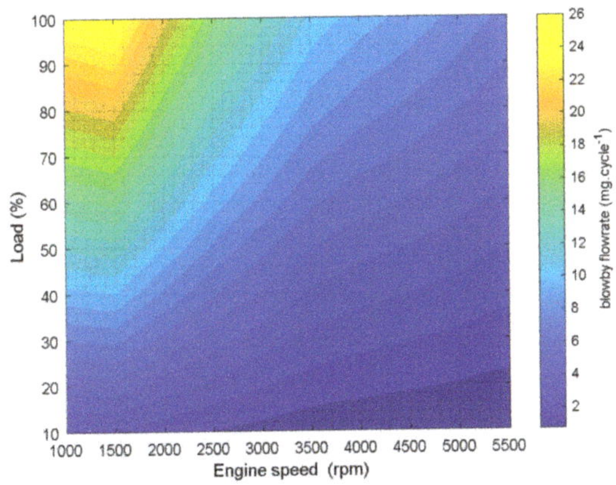

Figure 7. Blowby flowrate as a function of engine speed and load.

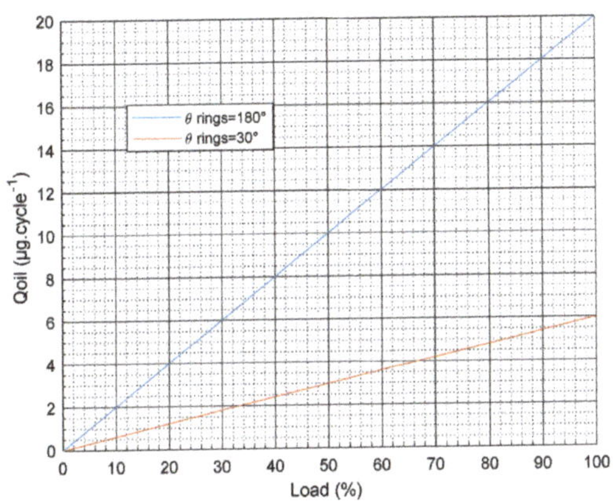

Figure 8. Quantity of oil swept by the blowby gases as a function of the endgap position at 1500 rpm.

Whereas if the endgaps are close, i.e., only 30° apart (cf., Figure 1b), the maximum amount of oil swept by the blowby gases can be at best 6 μg.cycle^{-1} for the same blowby flowrate. The same calculation can be made for each rotation speed, assuming an oil film height. It is therefore very clear that the oil concentration of the blowby gases is directly related to the position of the endgap and not only to its flowrate. The more oil is swept by the blowby gases, the less oil is returned to the cylinder via the backflow gases. This variable phenomenon linked to the rotation of the rings is difficult to predict and is the reason for the variations in particle emissions linked to the oxidation of the oil on strictly identical tests [49].

3.3. Influence of Blowby on Particulate Emissions at Idle

At no load, i.e., when decelerating or idling, the pressure inside the cylinder is very low. The maximum pressure varies between 4 bars and 10 bars. Therefore, the blowby flowrate is very low and the oil sweeping phenomenon described by Thirouard, et al., [26]

(Equation (1)) is minimal. In Figure 9, the pressures between rings and the ring positions evolving at idle speed over a complete cycle are shown from the simulation. The upper graph represents the evolution of the pressures above the compression ring and below. The evolution of the pressures between the oil seal and oil scraper rings visible in the middle graph is approximately equal to the pressure in the oil pan, i.e., equal to 0.98 bar. Compared to the graph in Figure 5 at 2000 rpm and full load, the difference is very significant. The maximum pressure reaches 8 bars for Land 1 and 3.5 bars for Land 2. The lower graph shows the evolution of the axial position of the first two rings in relation to their respective lower grooves. During the intake stroke and most of the compression stroke, the compression ring (Ring 1) stays on the upper axial part of its groove. At the end of compression and expansion strokes, the ring stays on the lower part of its groove. Then, at the end of expansion stroke and at the beginning of the expulsion of the exhaust gases, the ring oscillates axially to come into contact with the upper part of the groove and finally returns into contact with the lower part of the groove at around 540° CA and until the end of the cycle. The sealing ring (Ring 2) evolves with the same trend as the compression ring with an additional oscillation between 680° CA and 720° CA.

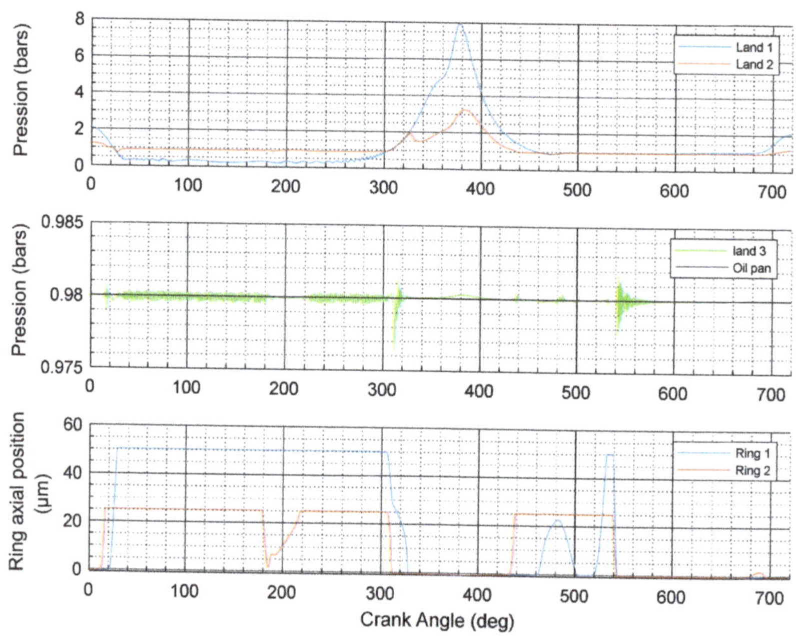

Figure 9. Evolution of pressures and axial positions of the piston rings at idle.

From above, it appears that the blowby flowrate seen in Figure 10 is very low and reaches its maximum of 0.28 g.s^{-1} at 320° CA and then stabilises at around 0.18 g.s^{-1} during the remainder of the compression stroke and the beginning of the piston expansion. During the expulsion of the exhaust gases, the blowby flowrate is close to 0. Then, due to the low pressure inside the cylinder during the intake stroke, the flowrate is negative and stabilises at around −0.15 g.s^{-1}. During this period, gases from the oil pan are redirected to the cylinder.

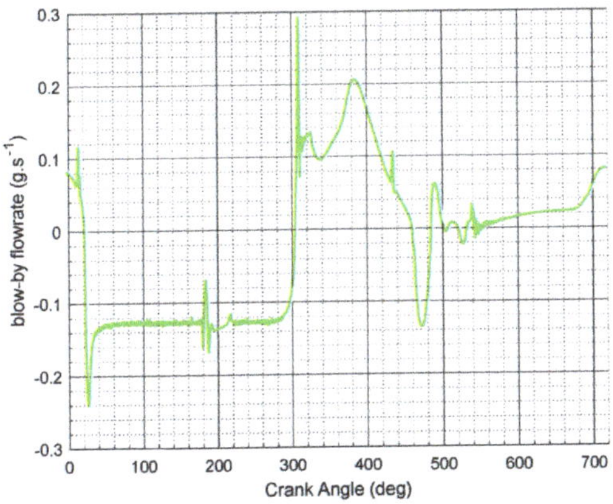

Figure 10. Evolution of blowby flowrate at idle.

In the end, the amount of blowby gas that passes through the piston to the crankcase is zero since this accumulation is negative and amounts to −3.4 mg.cycle^{-1}. This means that the phenomenon is reversed and that the gases from the oil pan could be directed towards the cylinder.

The oil sweeping effect is non-existent and the direction of the blowby flowrate tends to retain and push the oil stored in Lands 1 and 2 back into the cylinder. In addition, the compression and sealing rings oscillate two to three times per cycle within their respective grooves, and this creates a phenomenon of pumping and expulsion of the oil from the scraper ring to the crown of the piston via the sealing and compression rings. This phenomenon, described in particular by Thirouard [48] and Yilmaz, et al., [50] is significant at idle or low load.

Previous simulation results have shown that idling is conducive to an accumulation of oil at the piston crown. During rapid acceleration, the sudden increase in pressure and temperature in the cylinder will cause some of this oil to burn off, while some will be returned to the crankcase via the blowby phenomenon. The duration of the idling time between accelerations could have an impact on the amount of oil stored in the piston crown and consequently on oil consumption and particulate emissions. It is therefore important to study the impact of the duration of the idle time between two accelerations in order to check whether this accumulation phenomenon revealed by our simulation model is real. This is the subject of the following section.

4. Influence of Idling Time between Two Accelerations on Particulate Emissions

4.1. Setup

The tests were carried out on the engine mounted on a Horiba bench described in Section 2.1. With this bench, it was possible to create some transients. In order to be close to the real conditions of the "stop and start" system, an acceleration from idle to 2000 rpm and 60 Nm in 5 s was selected. To reach this point, it was necessary to keep the accelerator pedal pressed at 100% for about 2 s (part a, on the lower graph in Figure 11). Then, the engine stayed at 2000 rpm and 60 Nm during 15 s (part b) before slowing down (part c) and coming back in idle position (part d).

Three configurations were tested. For case 1, there was no idling time between two accelerations. The engine slowed down to 750 rpm before accelerating for a new transient. Case 2 consisted of an idling time of 7 s between two accelerations, while case 3 consisted of an idling time of 22 s. The test conditions are presented in the Table 3 below.

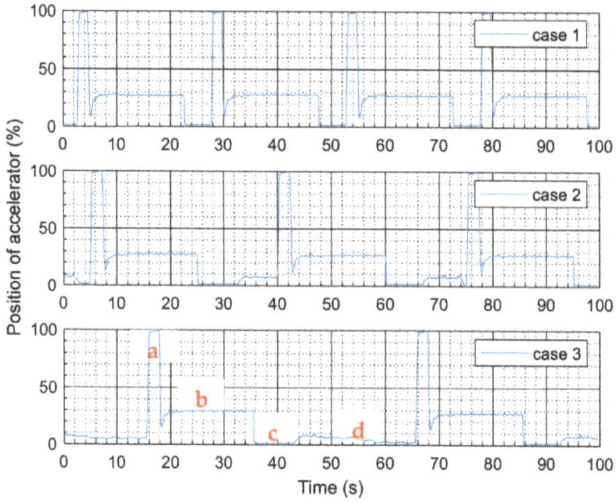

Figure 11. Cases description.

Table 3. Test conditions.

Pressure (mbars)		Averaged Temperature (°C)					
Atmospheric	Sample line PPS	Intake	Cooling	Oil	Engine bench	Exhaust	Sample line PPS
1013	1538	26	93	96	17	356	170

Figure 12 shows the evolution of engine speed, effective dynamic torque, in-cylinder pressure, and intake manifold pressure, depending on the position of the accelerator pedal during this transient.

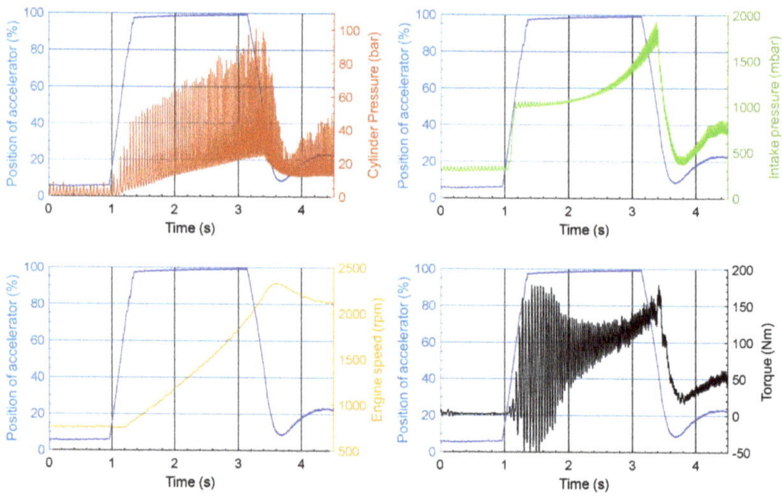

Figure 12. Evolution of engine parameters during a transient.

The signal "accelerator position" represents practically the position of the engine throttle, which means that it is in the fully open position for about 2 s before reaching the requested point. In Figure 12 bottom right, the dynamic effective torque during this transient is presented. The significant variations at the start are consequences of the important variations of the in-cylinder pressures between the cylinders. For example, for one cylinder, the in-cylinder pressure may be 25 bars while it may still be 6 bars for another cylinder (which is the idle in-cylinder pressure). This difference will generate a significant variation of the engine torque.

Berthome, et al., [13] exposed that particulate emissions vary greatly from test to test despite mastering both engine and bench parameters. It is, therefore, necessary to accumulate data and to proceed with a statistician approach to get a trend out, and thus evaluate the impact of the idling time between two transients. To obtain as much data as possible, 100 accelerations (strictly identical from the point of view of the control and test conditions) are carried out successively after reaching thermal stability and respecting a time of idling between each test for each configuration. The demonstration of the similarity of these transients is not the subject of this article and has already been validated in a previous article by Berthome, et al., [13,20].

4.2. Particles Measuring Device

The particle measurement device used in this study was a "Pegasor Particles Sensor" called a PPS, see Figure 2. The sample was not diluted, and this device measured the current carried out by the particles. For that, the PPS uses the corona effect to ionize particles: a corona discharge occurs when an electric current carried by a high voltage passes through two electrodes separated by a neutral gas (air).

This electric arc creates positive ions that propagate towards the molecules of air. This is called the ionization phenomenon. Then the ionized air mixes with the exhaust gases containing particles. These particles absorb the positive ions. The mixture is thus composed of free ions and ionized particles. An electric trap captures the free ions by maintaining a positive voltage on its surface. This has the effect of pushing the positive ions towards the external surface of the trap in order to discharge them. If the trapping voltage increases, then the lighter ionized particles are absorbed by the trap. It is possible to choose a sample of particles according to their size thanks to the choice of the trapping voltage. It can vary from 0 V to 1000 V according to the manufacturer data [51]. A trapping voltage of 400 V was used to measure particles with a size above or equal to 23 nm. This trapping voltage was selected because it corresponds to the minimum size required by the regulatory standards [1]. This technology is very reactive because it performs measurements at the frequency of 100 Hz. However, it is necessary to make the assumption of mean diameter, standard deviation, and fractal dimension in order to determine the number and the mass flow rate of the particles [52,53]. As particle emissions are higher during the transient phases, this study is performed in such an operating mode. However, it is not possible to know precisely the mean diameter of the particles, since it varies depending on the engine load and rotation speed.

Therefore, in this article, a simple comparison of the current measured by the PPS was made, because it is proportional to the particles emitted by the engine. This solution was chosen because the acquisition frequency of the PPS is high and allows the measurement of particle emissions during transients. Since a comparison will be made between the three cases, a relative measurement is sufficient.

4.3. Test Results and Analysis

Figure 13 shows the evolution of the average current carried by the particles of 100 strictly identical transients and for each case. The blue dash dotted-line curve represents the position of the accelerator pedal. At idle, the level of particulate emissions generated by the engine is very low and amounts to about 1 pA. Then, during acceleration, the current level rises sharply. In fact, to obtain rapid acceleration, it is necessary to increase the equivalence ratio to counteract friction and especially inertial effects. These rapid variations of equivalence ratio generate many particles reaching a maximum around 2.2 s. Then, from 2.5 s, the acceleration slows down, which favours stoichiometric conditions and therefore generates fewer particles, and the current measured by the PPS decreases. Cases one and two have almost the same trends. The current peak reached 112 pA for case 1 and 149 pA for case 2. However, case 3 is more important, and the current peak reached around 415 pA.

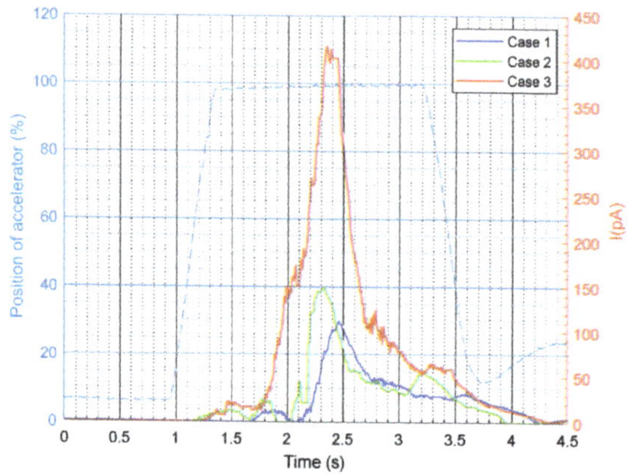

Figure 13. Evolution of particle emissions during the transient for each case.

The sum of current averaged for 100 transients ($\overline{\mathbf{IntegI}}$) for the three cases are shown below, in Table 4. The percentage standard deviation (σ) is calculated from the standard deviation of $\overline{\mathbf{IntegI}}$ of 100 transients divided by the mean. For case 1, i.e., without an idling phase, but only a deceleration phase between two transients (see part c Figure 11), the sum of the current averaged per 100 transients is 105 pA. For case 2, with an idling phase of 7 s between two transients, the sum of the current averaged per 100 transients is 133 pA, an increase of about 126% compared to case 1. For case 3, with an idling phase of 22 s between two transients, the sum of the current averaged per 100 transients is 329 pA, an increase of about 313% compared to case 1. The deviations between the three cases are very large even though the transients are strictly identical (cf part 4-2). Since it is always the same transients, the deviations in richness between each test remain very small and are not the cause of these differences. The only fluctuating phenomenon is the idling time between each acceleration. Consequently, it is the absence of oil sweeping by the blowby gases and its duration that leads to an accumulation of oil towards the crown of the piston. This oil is then projected towards the cylinder and is oxidised during strong acceleration, causing a peak in particle emissions. Therefore, the oil accumulation on the piston crown is proportional to the idling time between two accelerations. These experimental tests confirm the results put forward by the blowby simulation model.

Table 4. Results.

Case Number	Idling Time (s)	$\overline{\mathbf{IntegI}}$ (pA)	σ (%)
1	0	105	13
2	7	133	15
3	22	329	16

5. Conclusions

The creation of a blowby simulation model based on a TGDI engine made it possible to characterise the quantity of oil swept by these gases. This was performed at different engine operating points and as a function of the endgap position of both the compression and the sealing rings. Under full load at 2000 rpm, the quantity of oil swept can be as high as 20 μg. cycle^{-1}. At idle speed, the sweeping intensity is close to zero and there is a phenomenon of oil storage on the piston crown. The experimental results confirmed this observation and showed that the quantity of oil stored on the piston crown depended on the idling time. For example, the evolution of the idling time from 7 s to 22 s between two strictly identical accelerations generated three times more particles. The new Euro 7 standard should limit particulate emissions to 3 mg.km^{-1} and 6×10^{11} particles.km^{-1} on a WLTP cycle for particles >10 nm, instead of 23 nm previously. As there are eight idling stages with a total duration of approximately 224 s in a WLTP cycle, this phenomenon must be taken into account. At present, it is not possible to differentiate the oil pumping mechanism by the piston rings from the oil sweeping mechanism by the blowby gases at idle. A future study on a single-cylinder spark ignition engine will characterise the impact of each mechanism.

Author Contributions: Conceptualization, V.B. and D.C.; formal analysis, V.B.; investigation, V.B.; methodology, V.B.; project administration, V.B., D.C. and J.-F.H.; resources, D.C. and J.-F.H.; software, V.B.; supervision, V.B., D.C. and J.-F.H.; validation, V.B., D.C. and J.-F.H.; visualization, V.B., D.C. and J.-F.H.; writing—original draft, V.B.; writing—review and editing, V.B., D.C. and J.-F.H. All authors have read and agreed to the published version of the manuscript.

Funding: This research received no external funding.

Data Availability Statement: Not applicable.

Conflicts of Interest: The authors declare no conflict of interest.

Nomenclature

m	Kg	Ring mass
h	m	Variation of axial ring position in the groove
F_p	N	Pressure forces
F_i	N	Inertial forces
F_{fr}	N	Friction forces
F_{oil}	N	Oil film pressure forces
F_a	N	Adhesion forces
Q_{blowby}	m^3.cycle^{-1}	Blowby flow
μ_{air}	Pa.s	Dynamic viscosity of air
μ_{oil}	Pa.s	Dynamic viscosity of oil
θ_{rings}	°	Angle formed between the endgap of the first two rings
R_{land}	m	Clearance between piston radius and cylinder radius
h_{air}	m	Clearance between piston and cylinder
K_1 and K_2	-	Constants mainly related to the engine speed
WLTP	-	Worldwide Harmonized Light Vehicle Test Procedure
PAH	-	Polycyclic Aromatic Hydrocarbon
TGDI	-	Turbocharged Gasoline Direct Injection
PPS	-	Pegasor Particles Sensor
TDC	-	Top Dead Center
BDC	-	Bottom Dead Center
ICE	-	Internal Combustion Engine
Land 1	-	Space between the piston top and the first ring
Land 2	-	Space between the first two rings
Land 3	-	Space between the second and third ring
$\overline{\text{IntegI}}$	pA	$\frac{1}{n}\sum_{1}^{n}\int IdI$, avec n = test number
σ	-	Standard deviation

References

1. Commission Européenne. Règlement (UE) n°459/2012. *J. Off. Union Eur.* **2012**, *2012*, 16–24.
2. Barbussse, S.; Plassat, G. Les Particules de Combustion Automobile et Leurs Dispositifs D'élimination. 2005. Available online: http://scholar.google.com/scholar?hl=en&btnG=Search&q=intitle:Les+particules+de+combustion+automobile+et+leurs+dispositifs+d??limination#0 (accessed on 5 October 2021).

3. Marchal, C. Modélisation de la Formation et de L'oxydation des Suies Dans un Moteur Automobile. Ph.D. Thesis, Université d'Orléans, Orléans, France, 2008.
4. Law, M.E.; Westmoreland, P.; Cool, T.A.; Wang, J.; Hansen, N.; Taatjes, C.; Kasper, T. Benzene precursors and formation routes in a stoichiometric cyclohexane flame. *Proc. Combust. Inst.* **2007**, *31*, 565–573. [CrossRef]
5. Ketterer, J.E. Soot Formation in Direct Injection Spark Ignition Engines Under Cold-Idle Operating Conditions. Ph.D. Thesis, Massachusetts Institute of Technology, Cambridge, MA, USA, 2013.
6. Krestinin, A. Detailed modeling of soot formation in hydrocarbon pyrolysis. *Combust. Flame* **2000**, *121*, 513–524. [CrossRef]
7. Bockhorn, H. Detailed Mechanism and Modeling of Soot Particle Formation. In *Soot Formation in Combustion*; Springer: Berlin/Heidelberg, Germany, 1994.
8. Kittelson, D.B. Engines and nanoparticles: A review. *J. Aerosol Sci.* **1998**, *29*, 575–588. [CrossRef]
9. Cheng, W.K.; Hamrin, D.; Heywood, J.B.; Hochgreb, S.; Min, K.; Norris, M. *An Overview of Hydrocarbon Emissions Mechanisms in Spark-Ignition Engines*; Technical Papers; SAE International: Warrendale, PA, USA, 1993. [CrossRef]
10. Thawko, A.; Yadav, H.; Eyal, A.; Shapiro, M.; Tartakovsky, L. Particle emissions of direct injection internal combustion engine fed with a hydrogen-rich reformate. *Int. J. Hydrog. Energy* **2019**, *44*, 28342–28356. [CrossRef]
11. Swanson, J.; Ragatz, A.; Watts, W.; Kittelson, D.; Winsor, R. Nanoparticle measurements used to detect an engine oil control ring failure. *Proc. Inst. Mech. Eng. Part D J. Automob. Eng.* **2009**, *223*, 1071–1076. [CrossRef]
12. Berthome, V. Etude des Émissions de Particules d'un Moteur à Combustion Interne de Type Allumage Commandé. Ph.D. Thesis, Ecole Centrale de Nantes, Nantes, France, 2021.
13. Berthome, V.; Chalet, D.; Hetet, J.-F. Characterization of Particle Emissions of Turbocharged Direct Injection Gasoline Engine in Transients and Hot Start Conditions. *J. Therm. Sci.* **2021**, *30*, 2056–2070. [CrossRef]
14. Amirante, R.; Distaso, E.; Tamburrano, P.; Reitz, R.D. *Measured and Predicted Soot Particle Emissions from Natural Gas Engines*; Technical Papers; SAE International: Warrendale, PA, USA, 2015; Volume 2015. [CrossRef]
15. Delvigne, T. Oil Consumption Sources in a Modern Gasoline Engine Including Contribution of Blow-by Separator and Turbocharger: An Experimental Study Based on the Use of Radiotracers. *SAE Int. J. Fuels Lubr.* **2010**, 3, 916–924. [CrossRef]
16. Manni, M.; Carriero, M.; Roselli, A. *A Study of Oil Consumption on a Diesel Engine with Independently Lubricated Turbocharger*; Technical Papers; SAE International: Warrendale, PA, USA, 2002. [CrossRef]
17. Yilmaz, E. Sources and Characteristics of Oil Consumption in a Spark-Ignition Engine. Ph.D. Thesis, Massachusetts Instititute of Technologie, Cambridge, MA, USA, 2003.
18. Gohl, M.; Friedmann, M.; Holzmüller, J.; Wulff, R.; Braun, M. Impact of Lubricating Oil on Particle Emissions. In Proceedings of the SIA Powertrain & Power Electronics Congress, Port-Marly, France, 9–10 June 2021; pp. 47–54.
19. Yilmaz, E.; Tian, T.; Wong, V.W.; Heywood, J.B. *The Contribution of Different Oil Consumption Sources to Total Oil Consumption in a Spark Ignition Engine*; Technical Papers; SAE International: Warrendale, PA, USA, 2004. [CrossRef]
20. Berthome, V.; Chalet, D.; Hetet, J.-F. Impact of Blow-By Gas and Endgap Ring Position on the Variations of Particle Emissions in Gasoline Engines. *Energies* **2021**, *14*, 7492. [CrossRef]
21. Premnath, M.S.V.; Khalek, I.; Michlberger, A. *Effect of Lubricant Oil on Particle Emissions from a Gasoline Direct Injection Light-Duty Vehicle*; Technical Papers; SAE International: Warrendale, PA, USA, 2018; pp. 1–11. [CrossRef]
22. Amirante, R.; Distaso, E.; Napolitano, M.; Tamburrano, P.; Di Iorio, S.; Sementa, P.; Vaglieco, B.M.; Reitz, R.D. Effects of lubricant oil on particulate emissions from port-fuel and direct-injection spark-ignition engines. *Int. J. Engine Res.* **2017**, *18*, 606–620. [CrossRef]
23. Jues, T. Modélisation et Simulation des gaz de Blow-by dans un Décanteur Automobile. Ph.D Thesis, L'Ecole Nationale Supérieure d'Arts et Métiers, Paris, France, 2010.
24. Min, B.-S.; Kim, J.-S.; Oh, D.-Y.; Choi, J.-K.; Jin, J.-H. *Dynamic Characteristics of Oil Consumption- Relationship Between the Instantaneous Oil Consumption and the Location of Piston Ring Gap*; Technical Papers; SAE International: Warrendale, PA, USA, 1998. [CrossRef]
25. Gohl, M.; Matz, G.; Preuss, A.-C.; Pischinger, S.; Günther, M.; Ebert, T. *Investigation of Oil Sources in the Combustion Chamber of Direct Injection Gasoline Engines*; Technical Papers; SAE International: Warrendale, PA, USA, 2018; Volume 2018, pp. 1–11. [CrossRef]
26. Thirouard, B.; Hart, D.P. *Investigation of Oil Transport Mechanisms on the Piston Second Land of a Single Cylinder Diesel Engine, Using Two-Dimensional-Laser-Induzed Fluorescence*; Technical Papers; SAE International: Warrendale, PA, USA, 1998. [CrossRef]
27. Delprete, C.; Selmani, E.; Bisha, A. Gas escape to crankcase: Impact of system parameters on sealing behavior of a piston cylinder ring pack. *Int. J. Energy Environ. Eng.* **2019**, *10*, 207–220. [CrossRef]
28. Thirouard, B.; Tian, T. *Oil Transport in the Piston Ring Pack (Part II): Zone Analysis and Macro Oil Transport Model Benoist*; Technical Papers; SAE International: Warrendale, PA, USA, 2003. [CrossRef]
29. Nakashima, K.; Ishihara, S.; Urano, K. *Influence of Piston Ring Gaps on Lubricating Oil Flow into the Combustion Chamber*; Technical Papers; SAE International: Warrendale, PA, USA, 1995. [CrossRef]
30. Agarwal, V.K.; Ajay Paul, J. *Optimization of Piston and Ringpack Design to Improve the Performance and Emission Characteristics of a Gasoline Engine*; Technical Papers; SAE International: Warrendale, PA, USA, 2013; Volume 12. [CrossRef]
31. Tian, T. Dynamic behaviours of piston rings and their practical impact. Part 2: Oil transport, friction and wear of ring/liner interface and the effects of piston and ring dynamics. *Proc. Inst. Mech. Eng. Part J J. Eng. Tribol.* **2002**, *216*, 209–228. [CrossRef]

32. Iijima, N.; Miyamoto, T.; Takiguchi, M.; Kai, R.; Sato, M. *An Experimental Study on Phenomena of Piston Ring Collapse*; Technical Papers; SAE International: Warrendale, PA, USA, 2002. [CrossRef]
33. Tian, T. Dynamic behaviours of piston rings and their practical impact. Part 1: Ring flutter and ring collapse and their effects on gas flow and oil transport. *Proc. Inst. Mech. Eng. Part J J. Eng. Tribol.* **2002**, *216*, 209–228. [CrossRef]
34. Cheng, C.; Schock, H.; Richardson, D. The Dynamics of Second Ring Flutter and Collapse in Modern Diesel Engines. *J. Eng. Gas Turbines Power* **2015**, *137*, 111504. [CrossRef]
35. Rabute, R.; Tian, T. Challenges Involved in Piston Top Ring Designs for Modern SI Engines. *J. Eng. Gas Turbines Power* **2001**, *123*, 448–459. [CrossRef]
36. Wróblewski, P.; Koszalka, G. An Experimental Study on Frictional Losses of Coated Piston Rings with Symmetric and Asymmetric Geometry. *SAE Int. J. Engines* **2021**, *14*, 6. [CrossRef]
37. Wróblewski, P.; Iskra, A. *Problems of Reducing Friction Losses of a Piston-Ring-Cylinder Configuration in a Combustion Piston Engine with an Increased Isochoric Pressure Gain*; Technical Papers; SAE International: Warrendale, PA, USA, 2020. [CrossRef]
38. Zarenbin, V.; Kolesnikova, T.; Sakno, O.; Ollo, V.; Klimenko, V. Impact Evaluation of Piston Rings Mobility on a Gas Passage in an Internal Combustion Engine (ICE). *Sci. J. Sil. Univ. Technol. Ser. Transp.* **2019**, *104*, 187–201. [CrossRef]
39. Turnbull, R.; Dolatabadi, N.; Rahmani, R.; Rahnejat, H. An assessment of gas power leakage and frictional losses from the top compression ring of internal combustion engines. *Tribol. Int.* **2020**, *142*, 105991. [CrossRef]
40. Schneider, E.W.; Blossfeld, D.H. *Method for Measurement of Piston Ring Rotation in an Operating Engine*; Technical Papers; SAE International: Warrendale, PA, USA, 1990. [CrossRef]
41. Uhlig, B.P.; Kirner, C.; Preuss, A.-C.; Wachtmeister, G. *Real-Time Measurement of the Piston Ring Gap Positions and Their Effect on Exhaust Engine Oil Emission*; Technical Papers; SAE International: Warrendale, PA, USA, 2018; pp. 1–8. [CrossRef]
42. Thiel, N.; Weimar, H.-J.; Kamp, H.; Windisch, H. *Advanced Piston Cooling Efficiency: A Comparison of Different New Gallery Cooling Concepts*; Technical Papers; SAE International: Warrendale, PA, USA, 2007; Volume 2007, pp. 776–790. [CrossRef]
43. Abril, S.; García, C.; León, J. Numerical and Experimental Analysis of the Potential Fuel Savings and Reduction in CO Emissions by Implementing Cylinder Bore Coating Materials Applied to Diesel Engines. *Lubricants* **2021**, *9*, 19. [CrossRef]
44. Husberg, T.; Gjirja, S.; Denbratt, I.; Omrane, A.; Aldén, M.; Engström, J. *Piston Temperature Measurement by Use of Thermographic Phosphors and Thermocouples in a Heavy-Duty Diesel Engine Run Under Partly Premixed Conditions*; Technical Papers; SAE International: Warrendale, PA, USA, 2005. [CrossRef]
45. AVL GmbH. AVL 442, Operating and Service Manual Calibration and Procedure, AT0203E-07, November 1996, AVL List GmbH A8020 Gra-Austria. Available online: https://www.avl.com/documents/10138/2699442/AVL+Blow+By+Meter+Product+Description (accessed on 22 October 2022).
46. Koszalka, G.; Guzik, M. Mathematical Model of Piston Ring Sealing in Combustion Engine. *Pol. Marit. Res.* **2014**, *21*, 66–78. [CrossRef]
47. Takiguchi, M.; Nakayama, K.; Furuhama, S.; Yoshida, H. *Variation of Piston Ring Oil Film Thickness in an Internal Combustion Engine–Comparison Between Thrust and Anti-Thrust Sides*; Technical Papers; SAE International: Warrendale, PA, USA, 1998. [CrossRef]
48. Thirouard, B. Characterization and Modeling of the Fundamental Aspects of Oil Transport in the Piston Ring Pack of Internal Combustion Engines. Ph.D. Thesis, Massachusetts Institute of Technologie, Cambridge, MA, USA, 2001.
49. Tomizawa, K.; Ito, A. *Development of a New Method to Measure the Rotational Force Acting on the Piston Rings of a Gasoline Engine*; Technical Papers; SAE International: Warrendale, PA, USA, 2019; pp. 1–5. [CrossRef]
50. Yilmaz, E.; Thirouard, B.; Tian, T.; Wong, V.W.; Heywood, J.B.; Lee, N. *Analysis of Oil Consumption Behavior during Ramp Transients in a Production Spark Ignition Engine*; Technical Papers; SAE International: Warrendale, PA, USA, 2001. [CrossRef]
51. Pegasor. *User Manual*; Pegasor Ltd.: Tampere, Finland, 2014; Volume 3.
52. Amanatidis, S.; Maricq, M.M.; Ntziachristos, L.; Samaras, Z. Measuring number, mass, and size of exhaust particles with diffusion chargers: The dual Pegasor Particle Sensor. *J. Aerosol Sci.* **2016**, *92*, 1–15. [CrossRef]
53. Ntziachristos, L.; Amanatidis, S.; Samaras, Z.; Janka, K.; Tikkanen, J. Application of the Pegasor Particle Sensor for the Measurement of Mass and Particle Number Emissions. *SAE Int. J. Fuels Lubr.* **2013**, *6*, 521–531. [CrossRef]

energies MDPI

Article

Particle Size Distribution and Enrichment of Alkali and Heavy Metals in Fly Ash on Air and Oxy-Fuel Conditions from Sludge Combustion

Ha-Na Jang [1], Heung-Min Yoo [2] and Hang Seok Choi [1,*]

1 Department of Environmental Engineering, Yonsei University, Wonju 26493, Republic of Korea
2 Resource Recirculation Research Division, National Institute of Environmental Research, Inchon 22689, Republic of Korea
* Correspondence: hs.choi@yonsei.ac.kr; Tel.: +82-33-760-2485; Fax: +82-33-760-2571

Abstract: Comparative tests in air and oxy-fuel combustion were conducted in a 30 kW_{th} circulating fluidized bed (CFB) pilot plant for waste sludge combustion. General combustion characteristics of the CFB, such as pressure profiles, temperatures along the bed, and flue gas composition, were different under the air and oxy-fuel conditions. At the bottom and in the fly ash, alkali and heavy metals had different distributions under the air and oxy-fuel combustion conditions. The particle size distribution in fly ash from air combustion was dominated by coarse particles, over 2.5 μm in size, whereas with oxy-fuel combustion, most particles were submicron in size, approximately 0.1 μm, and a smaller quantity of coarse particles, over 2.5 μm in size, formed than with air combustion. Mass fractions of Al, Ca, and K, below 2.5 μm in size, were found in the ashes from oxy-fuel combustion and in higher quantity than those found in air combustion. Submicron particle formation from Cr, Ni, Cu, and Zn in the fly ash occurred more during oxy-fuel combustion than it did in air combustion.

Keywords: waste sewage sludge; oxy-fuel combustion; circulating fluidized bed; particle size distribution; alkali and heavy metals

Citation: Jang, H.-N.; Yoo, H.-M.; Choi, H.S. Particle Size Distribution and Enrichment of Alkali and Heavy Metals in Fly Ash on Air and Oxy-Fuel Conditions from Sludge Combustion. *Energies* **2023**, *16*, 145. https://doi.org/10.3390/en16010145

Academic Editors: Tomasz Czakiert and Monika Kosowska-Golachowska

Received: 7 October 2022
Revised: 14 November 2022
Accepted: 26 November 2022
Published: 23 December 2022

1. Introduction

Recently, the need for waste to be converted to energy has emerged globally. Waste sewage sludge is one of the renewable energy resources, and the amount of it has gradually increased over the years in Korea, where it reached up to 20 million tons in 2020. Since ocean dumping was prohibited after the international prohibition of waste sludge was announced in 2012, technology to convert waste sludge to energy has developed as an alternative disposal option.

The capacity of commercial fluidized bed combustion (FBC) plants for waste sludge is between 50 and 300 tons per day. These commercial plants combust waste sludge by utilizing air, generating much carbon dioxide, which is a greenhouse gas (GHG).

However, GHGs have become a worldwide issue, and converting waste, such as sludge, biomass, and municipal solid waste, to energy has been identified as a secondary source of GHG emissions, with fossil fuel combustion deemed a primary emission source. To mitigate global warming, carbon capture and storage (CCS) technology was developed to reduce GHGs, such as carbon dioxide, from anthropogenic emission sources. The effects of GHGs on global warming are acknowledged worldwide; therefore, GHG emission reduction has increased in importance.

Oxy-fuel combustion consumes a combination of oxygen greater than 95% in purity and recycles flue gas, which is enriched with carbon dioxide. During oxy-fuel combustion, a gas consisting primarily of carbon dioxide is generated that is ready for sequestration without stripping of the carbon dioxide from the flue gas. Due to the different surroundings during combustion, the flue gas composition of oxy-fuel combustion should be different than it is with air combustion [1].

In waste sludge combustion, the flue gas includes various air pollutants, such as ash, heavy metals, sulfur oxide, and nitrogen oxide. Among these air pollutants, the behaviors of ash and heavy metals are especially important because these compounds incur adverse health effects and prevent economical operation by causing fouling of, and deposition on, the boiler tube. There are numerous studies about the behavior of ash and heavy metals during conventional waste sludge combustion [1,2]. Rink et al. [3] studied the behavior of ash during sewage sludge combustion using a 300 kW bubbling fluidized bed (BFB) combustor. According to this study, particle size distribution was different along the height of the BFB combustor. This might have occurred because of different particle formation mechanisms in each temperature region. In addition, particle shapes were different at each sampling location, which were related to the particle formation mechanism of heavy metals. Cenni et al. [4] investigated the recovery rate of heavy metals in bottom ash, cyclone ash, and filter ash by blending sewage sludge with coal using a 500 kW pulverized fuel combustion chamber. This study indicated that the recovery rate of metals tended to increase when the blending rate of sewage sludge increased due to a flame temperature difference, in comparison with coal combustion. Latva-Somppi et al. [5] studied ash deposition and size distribution from the combustion of sewage sludge with wood by comparing BFB and CFB combustors. According to this study, sewage sludge contained a high proportion of alkali metals, such as Al, Ca, Si, and K, which caused ash deposition in the refractory liners and boiler tube. It was found that ash deposition did not occur in the CFB combustor, due to the advantageous heat distribution of the CFB. In addition, ash size in the CFB was smaller than it was in the BFB because the time for particle growth from metals, which was the main cause of ash deposition, was shorter than it was for the BFB. Lopez et al. [6] investigated metal partitioning in bottom ash and fly ash in accordance with their volatility with variation in sludge mixture with coal using a 90 kW BFB pilot plant. According to the study, Hg, Cd, Cu, and Pb in the fly ash increased more significantly in the mixture of sludge with coal than in the coal alone, whereas Cr, Ni, Mn, and Zn in the fly ash were less significant. In addition, Hg, Cd, Pb, and Zn increased more in the fly ash than in the bottom ash, whereas Mn, Cu, Ni, and Cr were not as significant, which was likely related to the volatility of the metals. Marani et al. [7] studied the enrichment factor of metals, such as Cd, Cr, Mn, Ni, Pb, Ti, and Zn, in cyclone ash and fly ash from sewage sludge combustion by chlorine content using a 250 kg/h CFB pilot plant. According to the study, the concentrations of those metals were enriched more in the fly ash than they were in the cyclone ash. Regarding the enrichment factor, Cd and Cr increased as chlorine content increased, whereas other metals were not as significant, likely due to the difference in the volatility of metals and formation of metal chloride compounds. Amand and Leckner [8] studied mass balance of trace metals from co-combustion of sludge, with coal or wood as the base fuel, by using a 12 MW_{th} CFB boiler. The study indicated that the trace metals in ash increased when wood was the base fuel as the sludge mixing rate increased, whereas this trend was not present when coal was the base fuel. Regarding the mass balance of trace metals in ash, volatile matter, such as Hg and Cd, was enriched in the fly ash from a second cyclone and bag filter, but non-volatile matter, such as Mn, was evenly enriched in the bottom ash and fly ash during co-combustion of coal and sludge. However, non-volatile matter was enriched in the finest fly ash during co-combustion of wood and sludge. Elled et al. [9,10] studied relative enrichment of volatile matter and non-volatile matter at different sampling points, such as the bed ash, second cyclone ash, and bag-filter ash, from co-combustion of wood and sludge. The study indicated that as the sludge mixing rate increased, the enrichment rate of volatile matter increased more in the fly ash than it did for the non-volatile matter in the fly ash. As mentioned above, there has been numerous research works related to the behavior of ash and heavy metals during FBC incineration. In summary, the studies concluded that the fate of trace elements was influenced by fuel type, combustion facility type, and operating conditions (temperature, pressure, oxidizing environment, and ash formation). However, these studies were conducted mainly under air combustion conditions and focused on co-combustion of sludge with coal or wood as the base fuel using BFB combustion technology [11–15].

CFB combustion technology has many advantages for heat recovery, so waste sewage sludge combustion should be retrofitted to, or newly commercialized for, CFB combustion facilities. In addition, a commercial FBC should be retrofitted to a carbon dioxide reduction facility. Only a few studies have been conducted on oxy-fuel combustion technology using CFB combustion technology for sludge combustion. When these technologies are applied to sludge combustion, combustion performances and the behavior of ash and heavy metals should be different under different combustion conditions. In this study, the particle size distributions composed of affluent metal components and the chemical reaction of aluminum, calcium, and potassium as alkali metals and chrome, copper, nickel, and zinc as tract metals from sludge combustion under the conditions of oxygen with nitrogen as the air combustion and oxygen with carbon dioxide as the oxy-fuel combustion were demonstrated, using a 30 kW CFB reactor.

2. Test Facility and Experimental Methods

2.1. Test Facility and Fuel Characteristics

Figure 1 shows a schematic stream of the test facility of the 30 kW CFB oxy-fuel reactor. The demonstration test was conducted in the facility, which consisted of a riser, a cyclone, a down-comer, and a loop-seal. The facility had a riser with an inner diameter of 0.15 m and a height of 6.4 m. The combustion temperature for the sludge fuel combustion was optimized at 800 °C. The feeding rate of the sludge was determined at 13 kg/h. Table 1 shows the summary of experimental conditions using the 30 kW CFB oxy-fuel reactor [16].

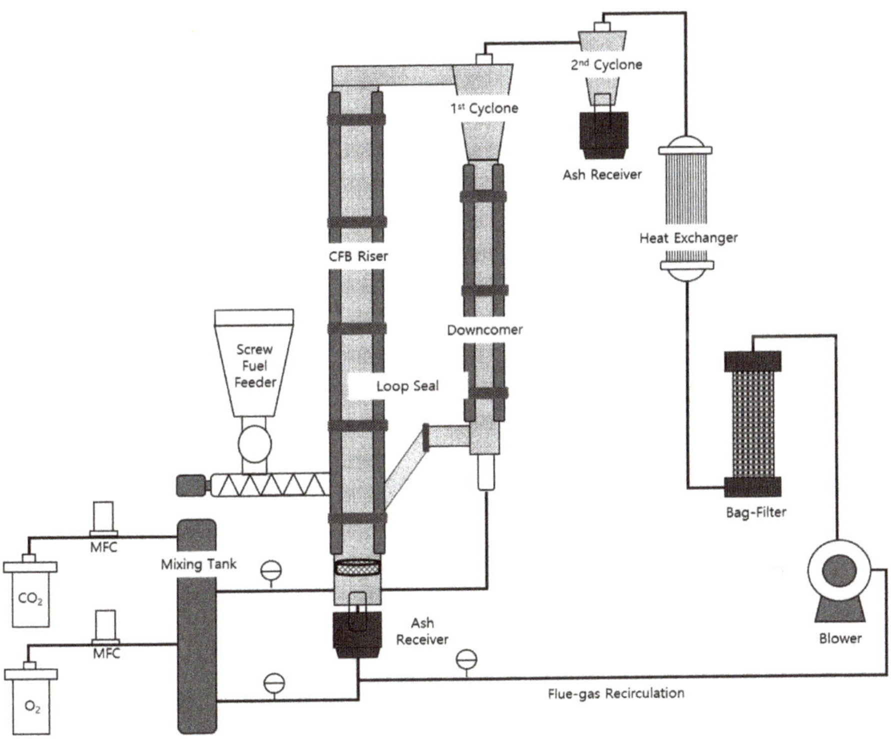

Figure 1. Schematic diagram of the 30 kW CFB oxy-fuel pilot test bed.

Table 1. Experimental conditions used during oxy-fuel combustion in the 30 kW CFB oxy-fuel pilot test bed.

Design Factor	Value
Bed diameter (m)	0.15
Fuel feeding rate (kg/h)	13
Solid fuel mixing rate (%)	0~30
Oxygen injection rate (%)	23
Combustion temperature (°C)	800
Flow rate (L/min)	900

2.2. Sampling and Analysis

The heating value of the sludge fuel was analyzed by an AC-350 calorimeter from the LECO Corporation in St. Joseph, MI, USA. The proximate analysis of sludge fuel was conducted by a TGA-601 from the LECO Corporation in MI, USA. The elemental analysis of sludge fuel was conducted by a 2400 series CHNS/O analyzer from Perkin Elmer Inc. in Waltham MA, USA. The analysis of the gaseous compounds in the outlet gas was conducted by PG-300 portable gas analyzer from HORIBA, Ltd in Kyoto, Japan. Particulate matter below 10 μm in size, produced from the conditions of oxygen with nitrogen and oxygen with carbon dioxide, was sampled by a low-pressure impactor from Dekati Ltd. in Kangasala, Finland. Sludge fuel and particle samples were pretreated by means of the EPA 3050B method, and metal compounds in the sludge fuel and fine particulate matter were analyzed by an ICP-MS from Gilson Inc. in Middleton, WI, USA.

3. Results and Discussion

3.1. Combustion Surroundings and Flue Gas Composition

The pressure gradient in the reactor was investigated under the conditions of oxygen with nitrogen and oxygen with carbon dioxide. Table 2 shows results of basic characteristic analysis of waste sewage sludge. In proximate analysis, waste sludge fuel contains much of volatiles and ashes. The calorific value was 3008 kcal/kg. In element analysis, carbon and hydrogen contents were 28.14 and 4.74, respectively. Chloride content was 530 ppm. In metal analysis, sludge fuel contained much of alkali metals and toxic heavy metals. Figure 2 shows the pressure and temperature profiles in the reactor under the conditions of oxygen with nitrogen and oxygen with carbon dioxide. As shown in Figure 2a, the pressure gradient showed a typical pressure trend of the CFB at each point under both conditions of oxygen with nitrogen and oxygen with carbon dioxide. The decrease in pressure under the oxygen with nitrogen condition was higher than that under the oxygen with carbon dioxide condition. Table 3 shows the gas components of fluidization air. In a previous study, the drop in pressure under the oxygen with carbon dioxide condition with fluidized air, that consisted of carbon dioxide and oxygen, increased as the oxygen rate increased from 21% to 40%, because the kinematic viscosity of the fluidizing injection gas increased as the oxygen rate increased. The kinematic viscosity under the oxygen with nitrogen condition was larger than under the oxygen with carbon dioxide condition. The drop in pressure increased as the kinematic viscosity of fluidizing injection gas increased. As shown in Figure 2b, the temperature gradient in the reactor was uniform as the axis of height under the conditions of oxygen with nitrogen and oxygen with carbon dioxide. In a previous study, the temperature gradient under the oxygen with nitrogen condition was relatively higher than under the 21% oxygen with carbon dioxide condition [9]. As shown in Table 3, the heat capacity of CO_2 was much larger than that of N_2. It caused the flame temperature of the 21% oxygen with carbon dioxide condition to be lower than that of the oxygen with nitrogen condition since the fuel sludge had much volatile materials. It caused delayed devolatilization and delay in the ignition time of sludge fuel under the 21% of oxygen with carbon dioxide condition because much of the carbon dioxide was utilized, instead of nitrogen, on FBC combustion. However, the temperature gradient under the 23% of oxygen with carbon dioxide condition was higher than that of the 21%

of oxygen with nitrogen condition. It was considered that the 23% of oxygen with carbon dioxide condition caused higher flame temperature than that of oxygen with nitrogen combustion, and devolatilization and total ignition time were fastly declined under the 23% oxygen condition. Table 4 shows the outlet gas temperature under the conditions of oxygen with nitrogen and oxygen with carbon dioxide. Regarding flue gas temperature, the temperature was relatively higher under the oxygen with carbon dioxide condition than that under the oxygen with nitrogen condition. This was due to the fact that H_2O and CO_2 in the flue gas had larger heat capacity than N_2 and O_2, as shown in Table 3, and the temperature of the outlet gas, including more H_2O and CO_2, was larger than it was for typical air conditions. Furthermore, it was indicated that the sludge fuel had much volatiles, and the heat distribution from sludge fuel combustion was reached further along the CFB combustion system under the oxygen with carbon dioxide condition than that under the oxygen with nitrogen condition.

Table 2. Results of basic characteristic analysis of waste sewage sludge.

Proximate Analysis (wt, %)		Element Analysis (wt, %)	
Moisture	7.32	Carbon	28.14
Volatile	45.11	Hydrogen	4.74
Fixed carbon	12.25	Nitrogen	4.43
Ash	35.04	Oxygen	23.90
-	-	Sulfur	0.43
Calorific value (kcal/kg)	3008	Chloride	0.053
Selected metals analysis			
Alkali metals analysis (ppm)		**Toxic heavy metals analysis (ppm)**	
Al	21,700.0	Zn	635.4
Ca	11,204.7	Cu	305.2
K	8249.0	Cr	42.5
-	-	Ni	30.8

Table 3. Physical gas properties of fluid flow utilized as fluidization air.

	H_2O	O_2	N_2	CO_2	Ratio, CO_2/N_2
Density (ρ) [kg/m^3]	0.157	0.278	0.244	0.383	1.6
Thermal conductivity (k) [W/m·k]	0.136	0.087	0.082	0.097	1.2
Specific heat capacity (c_p) [J/mol·°C]	45.67	36.08	34.18	57.83	1.7
Kinematic viscosity (m^2/s)	3.20	2.09×10^{-4}	2.00×10^{-4}	1.31×10^{-4}	0.7

Table 4. Flue gas composition and temperature during the air and oxy-fuel combustion.

Test Condition	O_2 (%)	CO_2 (%)	CO (%)	Temp (°C)
Air	5.1	15.3	0.9	697.4
Oxy-fuel	6.4	82.9	1.5	736.9

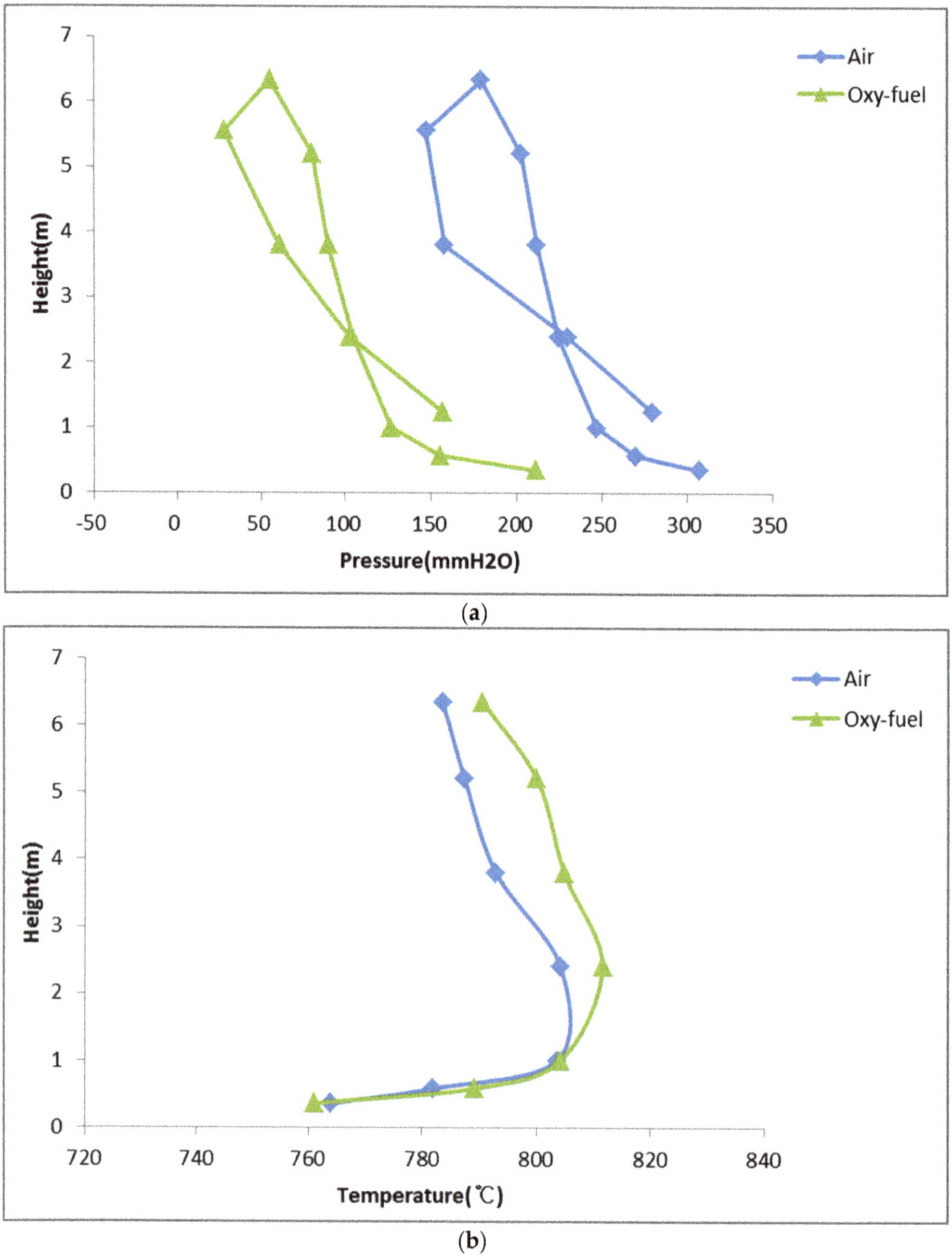

Figure 2. The pressure and temperature profile from air and oxy-fuel combustion of waste sewage sludge: (**a**) pressure profile (**b**) temperature profile.

3.2. The Behavior of Ash and Heavy Metals

(1) Metal Composition of Bottom and Fly Ash

Sewage sludge consists of many volatiles, metal elements, and ash fractions. During sewage sludge combustion, metal elements undergo a series of reaction mechanisms, such as volatilization, nucleation, condensation, and coagulation for particulate formation [2,17]. These particulate formation mechanisms in bottom ash and fly ash under the oxygen with carbon dioxide condition were affected by different conditions in the combustion environment. Figure 3a shows the comparative concentration of alkali metals in the bottom ash and fly ash under the conditions of oxygen with nitrogen and oxygen with carbon dioxide. The species of aluminum, calcium, and potassium were distributed mainly in the bottom and fly ash from sludge fuel combustion under the conditions of oxygen with nitrogen and oxygen with carbon dioxide. The metal components tended to be associated with particle formation mechanisms by causing the growth of fly ash from sludge fuel combustion. In sludge fuel combustion, agglomeration and fouling are highly involved with calcium and alkali compounds, such as the chloride and sulfate species, which block the distribution of fluidizing air and heat exchange. Aluminum compounds are concerned with Cl-related corrosion, which arises mainly from the boiler of the sludge fuel combustion process. The Cl-related corrosion problem of Al compounds creates a serious problem for boilers and doubles the repair and operation costs of the sludge fuel combustion process. [18,19]. In bottom ash, the concentrations of alkali metals, such as aluminum, calcium, and potassium, under the oxygen with carbon dioxide condition were less than that under the oxygen with nitrogen condition. In fly ash, these metals showed a similar trend to that regarding the bottom ash under the oxygen with carbon dioxide condition. As shown, the concentrations of these metals in fly ash and bottom ash under the oxygen with carbon dioxide condition were lower than they were under the oxygen with nitrogen condition. It was indicated that the oxy-fuel combustion mitigated agglomeration, fouling, and corrosion problems from sludge fuel combustion, and was economically beneficial in terms of long-term operation of a sludge fuel combustion facility. In addition, oxy-fuel combustion for sewage sludge could contribute to a shorter ignition time of the metals and moderate particle growth in the bottom ash, which would temper de-fluidization by agglomeration, fouling, and corrosion in the CFB boiler. Figure 3b shows the comparative concentrations of heavy metals in the bottom and fly ashes under the conditions of oxygen with nitrogen and oxygen with carbon dioxide. Chrome, nickel, copper and zinc were portioned mostly in the bottom and fly ashes from sludge combustion under the conditions of oxygen with nitrogen and oxygen with carbon dioxide. Copper and zinc compounds in bottom ash and fly ash from sludge fuel combustion were less under the oxygen with carbon dioxide condition than that under the oxygen with nitrogen condition. In the oxygen with carbon dioxide condition, the gas circumstances were significantly different than under the oxygen with nitrogen condition. Lots of CO_2 under oxygen with the carbon dioxide condition had larger heat capacity than N_2 under the oxygen with nitrogen condition. Accordingly, CO_2 tended to take more heat in the condition of oxygen with carbon dioxide than the condition of oxygen with nitrogen condition, which caused decrease in combustion flame temperature. Eventually, a lower adiabatic temperature for oxy-fuel combustion occurred and caused a delay in ignition time, which delayed particle growth and formation, which was the pathway for the nucleation, vaporization, condensation, and coagulation mechanisms of those metals. However, an ignition delay would be rapidly declined over 23% of oxygen injection with decreased carbon dioxide rate for oxy-fuel combustion. Due to decreased carbon dioxide injection rate, the adiabatic flame temperature of this condition would be higher than it was for the air and 21% of oxygen injection rate for oxy-fuel combustion. As shown figure, the concentrations of alkali and heavy metals showed different trends for air and oxy-fuel combustion. The concentration of alkali metals, such as aluminum, calcium, and potassium in the fly ash under this condition were lower than they were under air combustion. Regarding heavy metals, the concentrations of zinc and copper compounds in the fly ash under this condition were lower than they were from air combustion. Cr and Ni compounds in both ashes showed a different trend from the previously mentioned metals, with greater increase in ranges of oxy-fuel combustion than in air combustion. It was indicated that particle size distribution of fly ash should be changed along the air and oxy-fuel conditions because the different combustion surroundings and adiabatic flame temperatures likely affected the particle formation mechanisms from each metal compound in air and oxy-fuel conditions. Based on the test results, it was determined that oxy-fuel combustion was more efficient than air combustion in terms of heat recovery and economical operation by mitigating agglomeration, fouling, and corrosion problems from sewage sludge combustion.

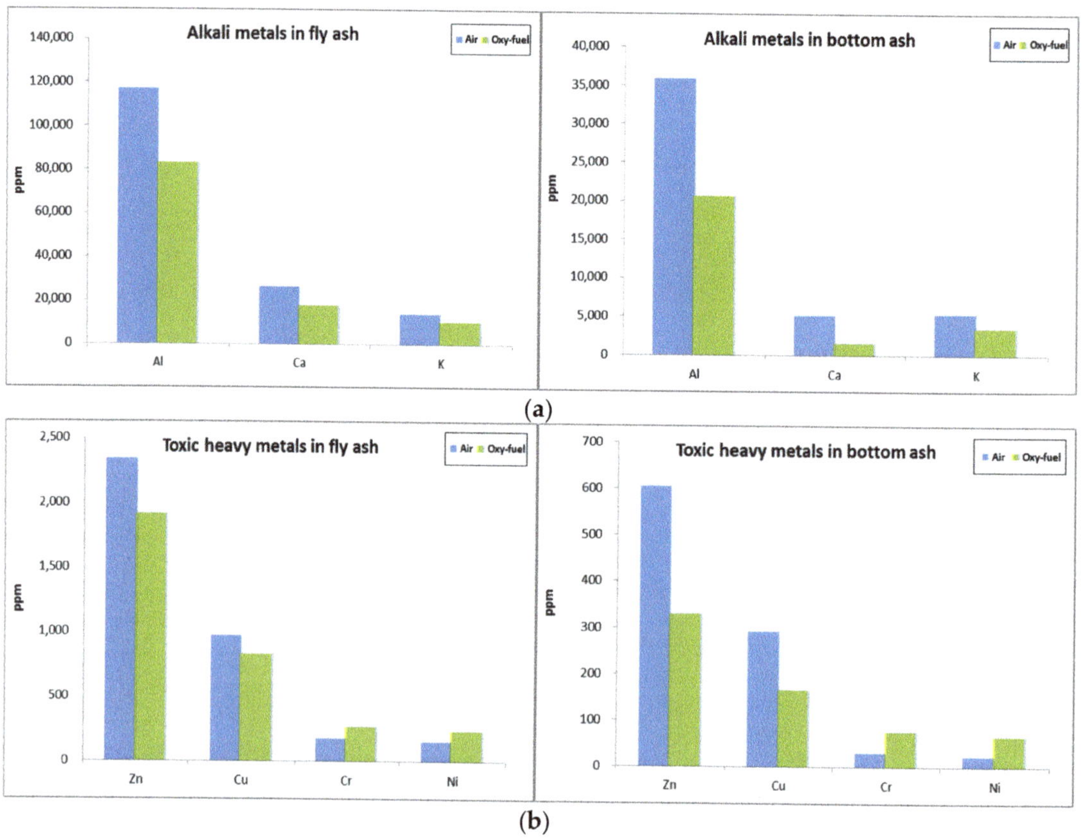

Figure 3. Concentrations of alkali and heavy metals in bottom and fly ash from air and oxy-fuel waste sewage sludge combustion: (**a**) alkali metals (**b**) heavy metals.

(2) Particle Size Distribution and Mass Fraction of Metals

As sewage sludge combustion generates a large amount of fly ash, an understanding of particle size distribution of fly ash is important because it mainly consists of alkali and heavy metals. It was hypothesized that particle size distribution would be affected by different surroundings during air and oxy-fuel combustion. Figure 4 shows the particle size distribution in fly ash from air and oxy-fuel combustion of sewage sludge. Particle size distribution from air combustion was mainly accumulated by coarse particles over 2.5 μm in size, whereas from oxy-fuel combustion, it showed each accumulation mode in fine particles below 1 μm in size, super-micron particles ranging in size from 1 μm to 2.5 μm, and coarse particles over 2.5 μm in size. Generally, fine particle formation was followed by a series of reactions of metal elements, such as nucleation, condensation, and coagulation. The sub-micron mode was depicted by each formation process, including direct vaporization of volatile metals and chemical reactions of refractory metal oxides. Refractory metal oxides (MO_X), in particular, could be reduced to sub-oxides (MO_{X-1}) by the reaction shown in Equation (1).

$$MO_X + CO \leftrightarrow MO_{X-1} + CO_2 \quad (1)$$

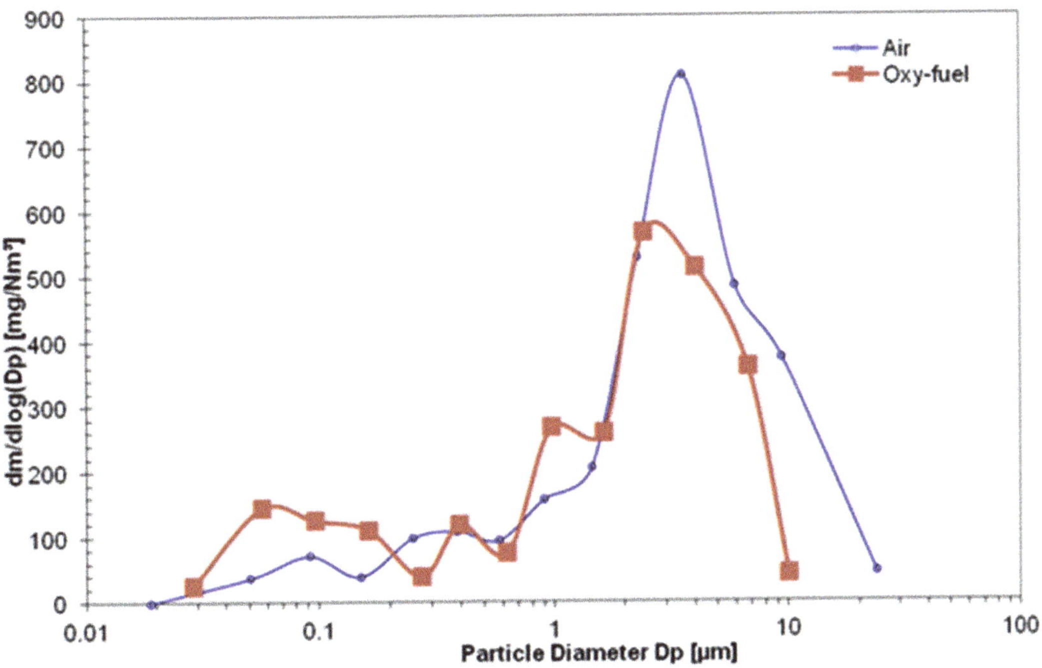

Figure 4. Particle size distribution in fly ash from air and oxy-fuel combustion of sewage sludge.

Sub-oxide metals were de-volatilized easily because of their low melting point, and then rapidly re-oxidized to gas phase to form fine particles, by a series of particle formation mechanisms, such as nucleation, volatilization, and condensation. The formation mechanism of a fine particle was dependent on fuel type, combustion temperature, and residence time. On the other hand, super-micron particles were formed from the following: (1) coalescence of the included mineral, which was not volatilized as a sub-oxide metal; (2) fine fragmentation of the excluded mineral; and (3) the solid to particle mechanism of inherent refractory metals that had high melting points. Coarse particles were formed from non-volatile mineral inclusions and char fragmentation. Figure 5 shows the mass fractions of the alkali and heavy metals mainly distributed in the fly ash from the combustion under the oxygen and nitrogen condition compared to the distribution under the oxygen and carbon dioxide condition. Aluminum, calcium and potassium were largely contained in the particle size between 1 µm and 2.5 µm under the conditions of oxygen with nitrogen and oxygen with carbon dioxide. In general, the size of the particles was directly generated from the metals inherent to sludge combustion. These metals were inherently refractory metals in fuel, and the particles were generated by the mechanisms of solid to particle, including those of inherently refractory metals. Fine particles were also generated by particle formation mechanisms, such as vaporization, nucleation, and condensation. The aluminum, calcium and potassium showed significant accumulation trend between 1 μm and 2.5 μm from the combustion of oxygen with nitrogen and oxygen with carbon dioxide. Generally, potassium tended to be much volatilized, which generated fine particles below 1 μm by particle generation mechanisms, such as vaporization and condensation. On the other hand, potassium also formed super-micron particles, larger than 1 μm. It was considered that metal vapor would spread on ash materials by chemical mechanisms, combine to aluminum and calcium, and finally cause conglomeration of particles. Under the oxygen with carbon dioxide combustion, the portion of aluminum, calcium and potassium were better conglomerated as fine particles below 1 μm than under oxygen with nitrogen combustion. The portions of aluminum, calcium and potassium under the oxygen with carbon dioxide condition, between 1 μm and 2.5 μm in size, were a bit larger than those under the oxygen with nitrogen condition, whereas the portion of aluminum, calcium and potassium under the oxygen with nitrogen condition, over 2.5 μm, were larger than those under the oxygen with carbon dioxide condition. This was explained by the fact that the ignition time delay under the oxygen with carbon dioxide condition was more rapid than that under the oxygen with nitrogen condition, and fine particle formation from aluminum, calcium and potassium was elevated by physical and chemical reactions. Fine particle formation from heavy metals, such as chrome, nickel, copper, and zinc, was conducted more intensively under the oxygen with carbon dioxide condition than that under the oxygen with nitrogen condition. It was considered that the fine particle formation from the oxygen with carbon dioxide condition was more intensive than that from the oxygen with nitrogen condition. The fine particles could be formed by the chemical reaction of the metal oxides. In general, the metal oxides could be changed into sub-oxide compounds when the original metal oxide reacted with carbon monoxide. The sub-oxide compounds were easily volatilized because of low melting temperature. Finally, because of the large amount of carbon dioxide, the vapor of the compounds could be more intensively re-oxidized to generate fine particles under the oxygen with carbon dioxide condition than those under the oxygen with nitrogen condition. Fine particle formation from zinc and copper would occur by volatilization and condensation, because the melting temperature of the compounds was lower than the adiabatic flame temperature under both conditions. On the other hand, the melting temperature of chrome and nickel was much higher. Finally, chrome and nickel mainly formulated in coarse particles, over 2.5 μm, under the oxygen and nitrogen condition.

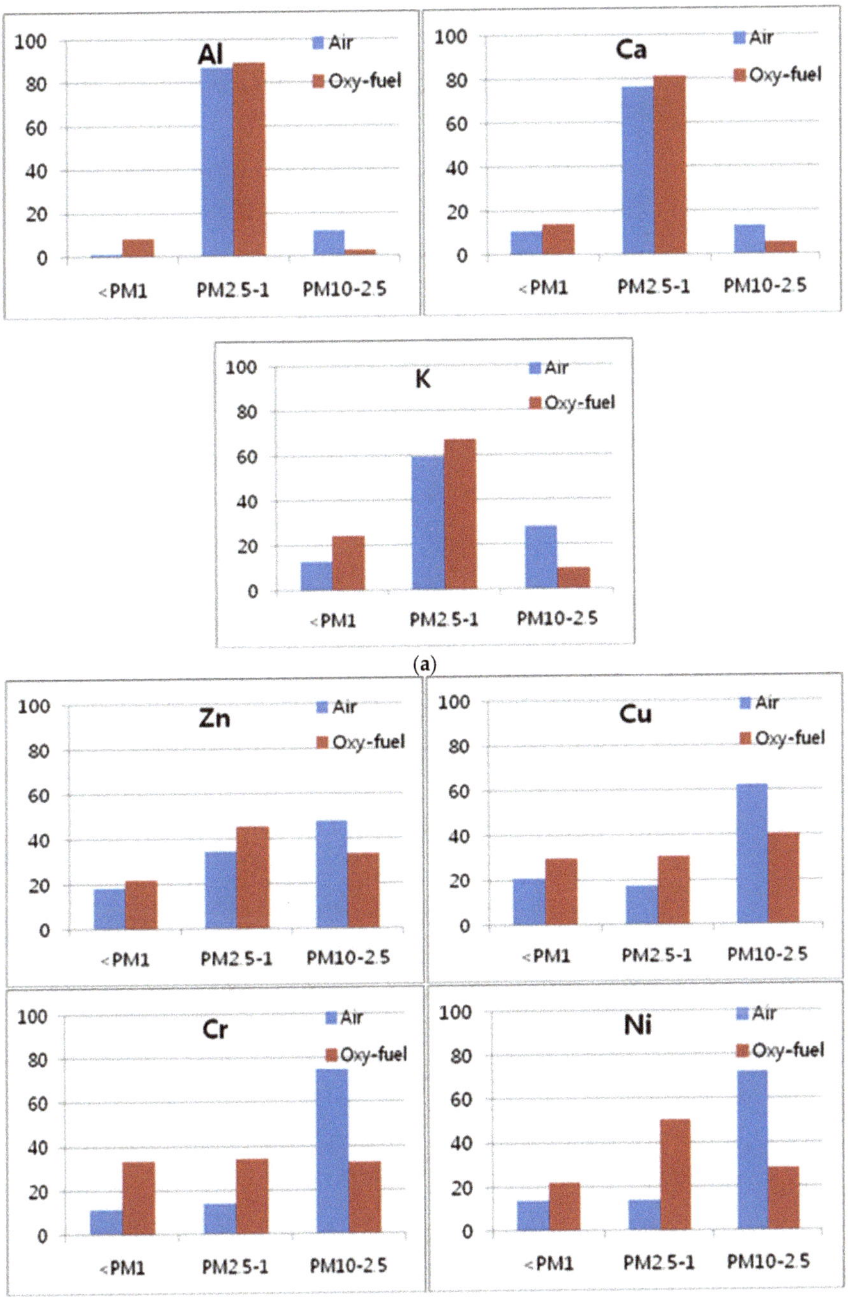

(b)

Figure 5. Mass fractions of alkali and heavy metals from air and oxy-fuel combustion: (**a**) alkali metals, (**b**) trace metals.

4. Conclusions

The comparative study on the behavior of alkali and heavy metals from waste sludge combustion under air and oxy-fuel conditions was conducted using a 30 kW CFB pilot test bed. Based on the experimental results, the major results are summarized as follows:

1. Temperature and pressure profiles in air and oxy-fuel combustion of sewage sludge were different according to mixtures of O_2/CO_2 and O_2/N_2. It was indicated that combustion surroundings for CFB waste sludge combustion changed due to physical gas properties, such as kinematic viscosity, density, and heat capacity of nitrogen, oxygen, and carbon dioxide during air and oxy-fuel combustion.
2. Based on flue gas and ash composition analysis in air and oxy-fuel combustion, the oxy-fuel combustion was more efficient than air combustion in terms of heat recovery, beneficial carbon dioxide capture, and economical long-term operation by mitigating agglomeration, fouling, and corrosion problems from sewage sludge combustion.
3. The PSD in fly ash under the oxygen with nitrogen condition was mainly distributed as coarse particles over 2.5 μm, whereas that under the oxygen with carbon dioxide plotted each peak mode as ultra-fine particle below 1 μm and fine particle between 1 μm and 2.5 μm. The results were caused by each series of mechanisms by metal compounds under different circumstances in both combustion conditions.
4. The portions of alkali metals under the oxygen with carbon dioxide condition were below 2.5 μm a bit larger than those under the oxygen with nitrogen condition. It is explained that the ignition time delay under the oxygen with carbon dioxide condition was more rapid than that under the oxygen with nitrogen condition. and fine particle formation from the metals was elevated by volatilization and condensation reactions.
5. Fine particle formation from chrome, nickel, copper, and zinc was more intensively conducted under the oxygen with carbon dioxide condition than under the oxygen with nitrogen condition. This was because of the large amount of carbon dioxide, and the compounds vapor could be more intensively re-oxidized to generate fine particles under the oxygen with carbon dioxide condition than those under the oxygen with nitrogen condition.

Author Contributions: Conceptualization, H.-N.J. and H.S.C.; methodology, H.-N.J.; software, H.-N.J.; validation, H.-N.J., H.-M.Y. and H.S.C.; formal analysis, H.-M.Y.; investigation, H.-M.Y.; resources, H.-M.Y.; data curation, H.-M.Y.; writing—original draft preparation, H.-N.J.; writing—review and editing, H.-N.J. and H.S.C.; visualization, H.-N.J.; supervision, H.S.C.; project administration, H.S.C.; funding acquisition, H.S.C. All authors have read and agreed to the published version of the manuscript.

Funding: This research received no external funding.

Acknowledgments: This work was supported by the Korea Ministry of Environment as the Field of Green Financing on Green Integration Human Resource Development Project. Also, This study was carried out with the support of 'R&D Program for Forest Science Technology (Project No. 2021356A00-2223-AC03)' provided by Korea Forest Service(Korea Forestry Promotion Institute).

Conflicts of Interest: The authors declare no conflict of interest.

References

1. Buhre, B.; Elliott, L.; Sheng, C.; Gupta, R.; Wall, T. Oxy-fuel combustion technology for coal-fired power generation. *Prog. Energy Combust. Sci.* **2005**, *31*, 283–307. [CrossRef]
2. Toftegaard, M.B.; Brix, J.; Jensen, P.A.; Glarborg, P.; Glarborg, A.D. Oxy-fuel combustion of solid fuels. *Prog. Energy Combust. Sci.* **2010**, *36*, 581–625. [CrossRef]
3. Rink, K.K.; Kozinski, J.A.; Lighty, J.S. Biosludge incineration in FBCs: Behavior of ash particles. *Combust. Flame* **1995**, *100*, 121–130. [CrossRef]
4. Cenni, R.; Frandsen, F.; Gerhardt, T.; Spliethoff, H.; Hein, K. Study on trace metal partitioning in pulverized combustion of bituminous coal and dry sewage sludge. *Waste Manag.* **1998**, *18*, 433–444. [CrossRef]
5. Latva-Somppi, J.; Kauppinen, E.; Valmari, T.; Ahonen, P.; Gurav, A.S.; Kodas, T.T.; Johanson, B. The ash formation during co-combustion of wood and sludge in industrial fluidized bed boilers. *Fuel Process. Technol.* **1998**, *54*, 79–94. [CrossRef]

6. Lopez, M.H.; Abelha, P.; Lapa, N.; Oliveira, J.S.; Cabrita, I.; Gulyurtlu, I. The behavior of ashes and heavy metals during the co-combustion of sewage sludge in a fluidized bed. *Waste Manag.* **2003**, *23*, 859–870.
7. Marani, D.; Braguglia, C.M.; Mininni, G.; Maccioni, F. Behaviour of Cd, Cr, Mn, Ni, Pb, and Zn in sewage sludge incineration by fluidized bed furnace. *Waste Manag.* **2003**, *23*, 117–124. [CrossRef] [PubMed]
8. Åmand, L.-E.; Leckner, B. Metal emissions from co-combustion of sewage sludge and coal/wood in fluidized bed. *Fuel* **2004**, *83*, 1803–1821. [CrossRef]
9. Elled, A.L.; Amand, L.E.; Leckner, B.; Andersson, B.A. Influence of phosphorus on sulphur capture during co-firing of sewage sludge with wood or bark in a fluidized bed. *Fuel* **2006**, *85*, 1671–1678. [CrossRef]
10. Elled, A.L.; Amand, L.E.; Leckner, B.; Andersson, B.A. The fate of trace elements in fluidized bed combustion of sewage sludge and wood. *Fuel* **2007**, *86*, 843–852. [CrossRef]
11. Van de Velden, M.; Dewil, R.; Baeyens, J.; Josson, L.; Lanssens, P. The distribution of heavy metals during fluidized bed combustion of sludge (FBSC). *J. Hazard. Mater.* **2008**, *151*, 96–102. [CrossRef]
12. Barbosa, R.; Lapa, N.; Boavida, D.; Lopes, H.; Gulyurtlu, I.; Mendes, B. Co-combustion of coal and sewage sludge: Chemical and ecotoxicological properties of ashes. *J. Hazard. Mater.* **2009**, *170*, 902–909. [CrossRef] [PubMed]
13. Solimene, R.; Urciuolo, M.; Cammarota, A.; Chirone, R.; Salatino, P.; Damonte, G.; Donati, C.; Puglisi, G. Devolatilization and ash comminution of two different sewage sludges under fluidized bed combustion conditions. *Exp. Therm. Fluid Sci.* **2010**, *34*, 387–395. [CrossRef]
14. Lopez, H.; Gulyurtlu, I.; Abelha, P.; Crujeira, T.; Salema, D.; Freire, M.; Pereira, R.; Cabrita, I. Particulate and PCDD/F emissions from coal co-firing with solid biofuels in a bubbling fluidized bed reactor. *Fuel* **2009**, *88*, 2373–2384. [CrossRef]
15. Urciuolo, M.; Solimene, R.; Chirone, R.; Salatino, P. Fluidized bed combustion and fragmentation of wet sewage sludge. *Exp. Therm. Fluid Sci.* **2012**, *43*, 97–104. [CrossRef]
16. Jang, H.-N.; Kim, J.-H.; Back, S.-K.; Sung, J.-H.; Yoo, H.-M.; Choi, H.S.; Seo, Y.-C. Combustion characteristics of waste sludge at air and oxy-fuel combustion conditions in a circulating fluidized bed reactor. *Fuel* **2015**, *170*, 92–99. [CrossRef]
17. Fryda, L.; Sobrino, C.; Cieplik, M.; van de Kamp, W. Study on ash deposition under oxyfuel combustion of coal/biomass blends. *Fuel* **2010**, *89*, 1889–1902. [CrossRef]
18. Wu, Y.; Wang, C.; Tan, Y.; Jia, L.; Anthony, E.J. Characterization of ashes from a 100 kWth pilot-scale circulating fluidized bed with oxy-fuel combustion. *Appl. Energy* **2011**, *88*, 2940–2948. [CrossRef]
19. Fryda, L.; Sobrino, C.; Glazer, M.; Bertrand, C.; Cieplik, M. Study of ash deposition during coal combustion under oxyfuel conditions. *Fuel* **2012**, *92*, 308–317. [CrossRef]

energies

MDPI

Article

Impact of Primary Air Separation in a Grate Furnace on the Resulting Combustion Products

Michał Kozioł [1,*] and Joachim Kozioł [2,†]

1 Department of Technologies and Installations for Waste Management, Faculty of Energy and Environmental Engineering, Silesian University of Technology, 44-100 Gliwice, Poland
2 Faculty of Civil Engineering, Architecture and Environmental Engineering, University of Zielona Góra, 65-516 Zielona Góra, Poland
* Correspondence: michal.koziol@polsl.pl
† Retired Employee.

Abstract: When burning fuel in grate furnaces, supplying the right amount of air to them is as important as the method of air supply. In a furnace with a fixed grate, the supply method of primary air is determined by the distribution of the supplied air stream over time, and in a furnace with a movable grate, the said method involves the distribution of the stream along the active length of the grate. The need to account for air distribution is attributable to complex processes that occur during the combustion process. The paper describes experimental studies aimed at determining the influence of the distribution of the supplied primary air on the emission of CO_2, CO, SO_2, NO_x, and on the content of combustible parts in the slag. In all cases, the total amount of primary air supplied to the process as well as other process control parameters was identical, and only the distribution of primary air was different. The paper proposes the use of a generalized function to describe the distribution of air, defined by its total demand and the relative time R that fuel remains on the grate until the maximum air stream is obtained. The quantity R was accepted at the value ranging from 1/6 to 2/3. With the rise of R, the emissions of CO_2, CO, and SO_2 increased by 53%, 125%, and 27%, respectively, and the emissions of NO_x and the share of combustibles in the slag decreased by 12% and 79%, respectively.

Keywords: grate furnaces; combustion process control; primary air distribution; co-combustion; emission of gaseous pollutants

Citation: Kozioł, M.; Kozioł, J. Impact of Primary Air Separation in a Grate Furnace on the Resulting Combustion Products. *Energies* **2023**, *16*, 1647. https://doi.org/10.3390/en16041647

Academic Editors: Monika Kosowska-Golachowska, Tomasz Czakiert and Andres Siirde

Received: 15 November 2022
Revised: 31 January 2023
Accepted: 3 February 2023
Published: 7 February 2023

1. Introduction

1.1. The Role of Combustion Processes in Grate Furnaces

A significant part of the combustion processes of coal [1–6] and biomass [7–13] is carried out in grate furnaces. In the case of the above-mentioned fuels, such furnaces are most often used in heat plants and combined heat and power plants with small and medium power [1–5,8].

The combustion of solids, which will probably be implemented for the longest time in most countries of the world, will involve the combustion of waste and waste-based fuels. In recent years, a significant increase in the installations for energy management of municipal waste have taken place in the EU and China among other places [14–16]. In the EU, in the years 2008–2018, the number of such installations increased by 60, and currently only in Poland there are about 100 installations at various stages of planning and implementation (with only 8 operating so far) [14]. In China, after 2000, there was a rapid increase in the number of municipal waste incineration installations, resulting in the highest combined efficiency in the world currently [15]. At the same time, it should be emphasized that, depending on the country, grate furnaces are installed in about 70% to 100% of municipal waste incineration installations (e.g., 100% in Brazil and Australia, over 85% in the EU, and almost 70% in China) [15–17].

1.2. Advantages and Disadvantages of Combustion Processes in Grate Furnaces

The main advantages of grate furnaces include flexibility in terms of the properties of burned fuels, high reliability, ease of use, relatively low investment and operating costs in the case of structures with lower and medium power fired with coal [1,8,18,19]. In the case of waste incineration, only operating costs are usually lower [15,17].

The disadvantages of grate furnaces include increased emissions of CO, benzoalfapyrene, C_xH_y compared with dust and fluidized furnaces, as well as a greater share of combustible parts in the slag [4–6,13,19–21]. The above-mentioned disadvantages principally do not apply to structurally advanced furnaces in thermal waste treatment installations.

Another disadvantage of grate furnaces involves temperature limitation of the process aimed at preventing the melting of mineral fractions of the fuel, which would lead to conglomeration and sticking of slag to the grate [8,18]. To a large extent, the above-mentioned disadvantages result in another disadvantage, i.e., relatively low energy efficiency of installations equipped with the discussed furnaces [3–6,8,18].

1.3. The Course of the Combustion Process in Selected Grate Furnaces

There are many constructions of grates and grate furnaces. The differences in the structures result mainly from the fact that grates and furnaces are adapted to the properties of the combusted fuel, to furnace efficiency and its functions (e.g., stove furnace or furnace in a water or steam boiler) [2,3,10,17,22,23].

The simplest structures of grate furnaces include furnaces with stationary grates and furnaces with flat movable grates. The first design is the traditional solution found in the furnaces of household stoves and low-power boilers. Such boilers are most often used by single-family houses, small farms, and small public utility buildings. The furnaces with flat movable grates constitute a difficult-to-estimate part of all movable furnaces, approximately hundreds of thousands [4–6,24]. The power of such furnaces ranges from a few to almost 100 MW_t.

During the combustion of solid fuel in a combustion chamber with a stationary or movable grate, a number of processes take place, such as heating and drying of the fuel, degassing and partial gasification of the fuel, combustion of the obtained degassing and gasification products, and the combustion of the obtained carbonizate [17,23,25,26].

In the case of furnaces with a stationary grate, these processes take place for each portion of fuel fed to the furnace, and they dominate successively. There are also possible situations of a simultaneous occurrence of two or more processes in different parts of a given batch of fuel. During fuel combustion on movable grates, these processes take place along the length of the grate. In this case, these processes often overlap (in a given longitudinal section of the grate and the fuel, at different heights of the fuel layer and above it, different processes take place).

The combustion process changes the amount and properties of fuel over time (in the case of stationary grates) or along the length of the grate (in the case of furnaces with a mechanical grate). In each of the mentioned processes, there is also a different, often very different, air demand. The largest stream of air is required for the combustion of degassing and gasification products, and much less is needed for the combustion of carbonizate. On the other hand, for the heating and drying processes of fuel, an air supply is not necessary. The optimal amount of air necessary to be delivered to each of the subprocesses is additionally influenced by the thickness of the fuel layer and its properties [2,10,17,24–28].

In the grate furnaces in question, the majority of the oxidant is supplied to the combustion chamber as primary air. This air is supplied under the grate. The stream of this air also has the additional task of cooling the grate and protecting it from damage by excessively high temperatures.

Therefore, the regulation of the primary air stream is one of the most important control parameters for grate furnaces.

In the case of stationary furnaces, primary air regulation is possible by changing the efficiency of the primary air fan. In the absence of a fan, the control is possible only by adjusting the size of the opening through which primary air is sucked into the combustion chamber.

In order to be able to control the amount of primary air along the length of the movable flat grate, its supply is controlled in a specific zone. Depending on the structure, the space under the grate is divided into several zones [3,6,24,25,28]. The structure of the grate and that of the combustion chamber enable the feeding process of various primary air streams to individual zones. Most frequently, this is done with sector-based air boxes.

The basic problem related to the regulation of the amount of air supplied to specific zones results from the lack of knowledge of the curve describing the best air distribution along the length of the grate for a given fuel and furnace and for its efficiency. In the literature [3,29], we can find only qualitative information on the change in air demand along the length of the grate, without any description allowing for its more precise characteristics.

Another problem arises from the difficulty of assessing whether the amount of air supplied to a given zone is optimal or close to this value. Most often, in the case of industrial boilers and furnaces, the measurement of flue gas composition (or only the measurement of O_2 and/or CO_2 concentrations) is carried out at a very limited number of points, and often only at one point. This point, in turn, is often located at the exhaust outlet from the furnace or even from the entire installation. In such a case, the measurement results allow only for the assessment of the correctness of the course of the entire process, and they do not provide information on the possibility of its quick improvement. Such improvement is only possible by changing the setting of the control parameters by a trial-and-error method. Taking into account the efficiency change of the installation and the properties of the combusted fuel implies the necessity of frequent changes of the control parameters and long-term operation of the furnace in conditions significantly different from rational parameters. The exceptions include furnaces built in municipal waste incineration installations. In that case, the number of checkpoints measuring the most important (in terms of the regulation of the combustion process) components of the exhaust gas (i.e., CO, CO_2, and/or O_2) and the temperature is relatively large. However, even in the case of such installations, due to the turbulence of the process and often large dimensions of the combustion chamber, the adjustment is difficult [6,11,25,26].

The distribution of primary air in grate furnaces is generally viewed as one of the most important control parameters, e.g., [2,3,5,10,17,22,23,27]. However, most often the problem of primary air stream distribution is considered in the context of numerical simulations of combustion processes, e.g., [30–38].

There are few studies demonstrating the impact of primary air distribution on the obtained gaseous emissions along the length of the grate based on measurements. The [25,39] present the results of research studies carried out in industrial facilities. Due to the conditions of the carried-out research, the number of measurement points in the research presented in [39] was limited to five. In addition, only one primary air separation was taken into consideration in the publication.

In conclusion, we can state that one of the most important problems associated with the operation of grate furnaces is the supply of a suitable air stream under the grate. The mentioned stream is variable: in time (fixed grate furnaces) or along the length of the grate (mechanical furnaces).

The main objective of the study is to demonstrate the influence of air distribution on the formation dynamics of gaseous products of the combustion process. The dynamics is understood as the change in the stream of the above products over time. The impact in question was demonstrated based on the results of experimental tests for the theoretically determined air distribution function. It is also the basic innovative element of the study. The knowledge of the dynamics makes it possible to determine the total gaseous emissions of combustion products. The work also determined the effect of air distribution on the share of combustible parts in the slag.

2. Materials and Methods

2.1. Theoretical Basis of the Conducted Research

2.1.1. Generalization of Air Distribution in Grate Furnaces

The course of the combustion process depends mainly on the initial composition of the fuel, its initial amount, the time fuel stays on the grate (measured from the beginning of the combustion process), and the distribution of air fed to the furnace. In the case of a furnace with a movable grate, the time of fuel residence on the grate, at a constant grate move speed, determines the location (position) of the fuel on the grate:

$$l = w\,t, \tag{1}$$

where:

l—fuel location on the grate, distance from the point of fuel supply onto the grate, m;
w—grate movement speed, m/min;
t—fuel-stay time on the grate, min.

In the literature [3,25,29], information can be found on the applied experimental distribution of combustion air supplied to the grate. The authors of the present study in [24] defined the following generalized function of air distribution in grate furnaces:

$$\dot{V}(x) = a\left(1 - \frac{x}{X}\right)x\,exp\left(b\frac{x}{X}\right) \tag{2}$$

where:

a—empirical coefficients, m^3/min^2 or m^3/m^2;
b—empirical coefficients, dimensionless quantity;
x, X—variable specifying the fuel condition on the grate and its maximum value;
$\dot{V}(x)$—air stream corresponding to the variable x;
with:

- in furnaces with a fixed grate, the following should be assumed: $x = t$; $X = T$, min,
- in a furnace with a movable grate: $x = l$; $X = L$, m,

where:

T—maximum fuel-stay time on the grate, min.,
L—active length of the movable grate, m.

The function $\dot{V}(x)$ should satisfy the following conditions: $\dot{V}(x = 0) = 0$; $\dot{V}(x = X) = 0$.

The form of the function is characterized by empirical coefficients. The numerical value of the factor b follows from the condition:

$$\frac{d\dot{V}(x)}{dx}/_{x=RX} = 0, \tag{3}$$

where:

R—relative value of the variable x which defines the maximum stream of supplied air, referenced to the value of X; a dimensionless quantity.

In turn, the numerical value of the coefficient "a" determines the relationship:

$$V_{ad} = \int_{x=0}^{x=X} \dot{V}(x)dx \tag{4}$$

where:

V_{ad}—total demand for combustion air resulting from the initial composition and amount of fuel and from the ratio of excess air, m^3.

Using Equations (3) and (4) we obtain:

$$b = \frac{1-2R}{R(R-1)}, \tag{5}$$

$$a = \frac{V_{ad}}{X^2 \cdot (Y_1 - Y_2)}, \quad (6)$$

where:

$$Y_1 = \frac{1}{b^2}[(b-1)\exp(b) + 1] \quad (7)$$

$$Y_2 = \left(\frac{1}{b} - \frac{2}{b^2} + \frac{2}{b^3}\right)\exp(b) - \frac{2}{b^3}, \quad (8)$$

When analyzing the system of Equations (2), (5) and (6), it can be seen that the $\dot{V}(x)$ curves of air distribution in grate furnaces are explicitly determined by the value of the variable $x = R$ and by the total combustion air demand V_{ad}.

2.1.2. Determination of Total Emissions of Gaseous Combustion Products

The calculation dependencies given in the further part of the study apply to furnaces with a fixed grate ($x = t$). They are also correct for furnaces with a mechanical grate, by taking into account Equation (1) and by substituting the variable l instead of the variable t.

The emission of gaseous products is determined from the dependence:

$$\dot{m}_{j,R}(t) = \dot{V}_{sp}(t) e_{j,R}(t) \approx \dot{V}(t) e_{j,R}(t) \quad (9)$$

where:

$\dot{m}_{j,R}(t)$—emission of the j-th gaseous product at time t for the R-variant of air separation stream, g/min or mg/min,

with: $j \in [CO_2, CO, SO_2, NO_x]$;

$\dot{V}_{sp}(t)$ —dry exhaust gas stream at time t for R-th variant of air separation, m^3/min;

$e_{j,R}(t)$—concentration of the j-th component in the dry exhaust gas, for the R-th variant of air separation at time t, g/m^3 or mg/m^3.

In view of the difficulties with measuring the flue gas flow, it was assumed: $\dot{V}_{sp}(t) = \dot{V}(t)$. This problem is discussed in more detail in Section 2.1.3.

The air distribution is zonal (see Section 2.4). The flow of air $\dot{V}_i$ in the i-th zone is constant but different between specific zones. The value of the air stream in the i-th zone is determined using Equation (2), yielding:

$$\dot{V}_{i,R} = \frac{1}{t_{i,k} - t_{i,p}} \int_{t_{i,p}}^{t_{i,k}} \dot{V}_R(t) dt, \quad (10)$$

where:

$t_{i,p}$, $t_{i,k}$—start and end value of time t limiting the i-th feed zone of air, min., $i = (1 \ldots I)$.

$\dot{V}_{i,R}$—air stream in the i-th delivery zone, for the R-th variant of its distribution, m^3/min.

Air stream is a quantity that controls the combustion process. Its value results from the distribution function (2). The compliance of the above air stream with the assumptions should be controlled by measurement.

The concentrations of the i-th components of exhaust gas $e_{j,R}(t)$ are also measured. Since the analyzer used to measure the concentrations shows average values in a specific time step, the emission stream of exhaust gas components can be determined from the following relation:

$$\dot{m}_{j,R,i,n} = \dot{V}_{i,R} e_{j,R,i,n}, \quad (11)$$

where:

$\dot{m}_{j,R,i,n}$ —average emission stream of the j-th product, with R-th air separation, in the n-th time step, g/min or mg/min;

$e_{j,R,i,n}$ —average concentration of the j-th exhaust component in the n-th time step, of the i-th zone and the R-th variant of air supply, g/m^3 or mg/m^3;

and:
$n = (1 \ldots N)$
where:
N—number of time steps in the i-th air supply zone.
The total emissions of gaseous combustion products can be determined from the formula:

$$m_{j,R} = \int_{t=0}^{t=T} \dot{m}_{j,R}(t)dt = \Delta t \sum_{n=1}^{n=N} \dot{m}_{j,R,k} = \Delta t \sum_{i=1}^{i=I} \dot{V}_{i,R} \sum_{n=1}^{n=N} e_{j,R,i,n}, \quad (12)$$

where:

$\dot{m}_{j,R}$—total emission of the j-th product, with R-th air separation, g or mg;

$\dot{m}_{j,R}(t)$—emission stream of the j-th product, with R-th air separation, at time t, g/min or mg/min;

Δt—time step, min.

2.1.3. Determination of the Relative Uncertainty of the Assessment of Gaseous Emissions of Combustion Products

The relative uncertainty of the assessment of the total emissions of gaseous products (exhaust components) can be derived from the formula:

$$\frac{\delta_{m_{j,R}}}{m_{j,R}} = \sqrt{\left(\frac{\delta_{\Delta t}}{\Delta t}\right)^2 + \left(\frac{\delta_{\dot{V}}}{\dot{V}}\right)^2 + \left(\sum_{i=1}^{i=I} N_i\right)\left(\frac{\delta_e}{e_m}\right)^2}, \quad (13)$$

where:

$\delta_{m_{j,R}}$ —absolute error of the total emission assessment of the j-th exhaust component for the R-th variant of air separation, g or mg;

$\delta_{\Delta t}$, $\delta_{\dot{V}}$, δ_e—absolute errors in the assessment of the time step, dry exhaust gas stream, exhaust gas component concentration, respectively: min, m^3/min, g/min or mg/min.

Taking into account the assumptions that the air stream and the dry flue gas stream are equal, as specified in Section 2.1.2, one should take into account the error of the method. Thus, the value $\frac{\delta_{\dot{V}}}{\dot{V}}$ in the Equation (10) can be determined from the formula:

$$\frac{\delta_{\dot{V}}}{\dot{V}} = \sqrt{\left(\frac{\delta_{ex}}{\dot{V}}\right)^2 + \left(\frac{\delta_m}{\dot{V}}\right)^2}, \quad (14)$$

where:

δ_{ex}, δ_m—absolute error of the determination method of dry flue gas stream and the measurement of air stream, m^3/min.

In order to determine the error of the method, stoichiometric calculations were carried out to determine the ratio α of the air stream to the dry flue gas stream, for exemplary fuels and exemplary values of excess air ratio λ. The obtained values are given in Table 1.

As the table shows, the error of the method for the analyzed exemplary solid fuels for $\lambda = 1.0$ is less than 4%, and it decreases with the increase of the excess air ratio. For the value $\lambda = 1.6$, it does not exceed 2.5%. For the conducted research, it was 1.6%.

The remaining uncertainty assessments are based on relative measurement errors. For the presented studies, they were less than 6%.

The relative uncertainty of the total emission of gaseous combustion products, determined according to the Equation (13), was less than 8%.

Table 1. Calculated values of α (ratio of air stream to dry exhaust gas stream) for the exemplary fuels and for the values λ = 1.0; 1.4; 1.6 (λ—excess air ratio).

Fuel Type	Calorific Value MJ/kg	α		
		λ = 1.0	λ = 1.4	λ = 1.6
Anthracite	31.4	1.025	1.017	1.015
Fat coal	31.8	1.037	1.027	1.023
Gas coal	31.4	1.038	1.027	1.023
Lean coal	31.4	1.037	1.026	1.023
Gas coal *	26.8	1.020	1.014	1.014
Dried peat	13.5	1.029	1.020	1.018
Lignite	9.6	1.033	1.024	1.021
Sewage sludge *	10.3	1.043	1.030	1.025
Mix * of coal + sludge	24.5	1.022	1.016	1.014

*—fuel used in the presented research.

2.1.4. Determination of the Share of Combustible Parts in the Slag

In order to determine the proportion of combustible parts in the slag, the slag was collected from the furnace chamber, from which, three samples weighing about 10 g were collected after it was homogenized. In the samples, the combustible parts were determined in accordance with PN-Z-15008-03:1993 [40]. The share of combustible parts was accepted as the average value of these determinations. The average value of these determinations was accepted as the share of combustible parts.

As to the share of combustible parts in the slag, the determination uncertainty of this value was 5%. This value was based on the determinations carried out in an additional series of 7 samples taken from the furnace chamber carried out for one of the tests. This value was defined as the quotient of the standard deviation related to the mean value.

2.2. Research Stand

The laboratory stand used in the research enables the simulation of combustion processes taking place in water boilers with a fixed grate and with a movable grate (e.g., belt grate). The diagram of the test stand shown in Figures 1 and 2 presents the view of its most important part. The main element of the stand is the combustion chamber consisting of two basic parts: the upper and the lower part. In the lower part, it is possible to regulate the heating temperature of the chamber up to 1200 °C. In the upper part, a water jacket is applied.

The lower part of the installation, due to the possibility of heating to a constant temperature, enables the simulation of the vault ignition (afterburning) of a grate boiler. The heated chamber walls also simulate the influence of the remaining parts of the real boiler on the combustion fuel sample. The lower part of the test installation is designed to operate at temperatures up to 1800 °C. The upper part of the boiler allows the exhaust gas to be cooled in a manner similar to that occurring in the upper part of the combustion chambers of water boilers.

In order to quickly load the fuel sample into the furnace chamber and to allow the fuel to be placed on the grate, the grate is placed on a movable bed. The bed is pushed inside after the lower chamber has been heated to the assumed temperature.

The combustion chamber is equipped with a number of measuring nozzles that allow for temperature measurements and gas sampling at various points. The installation is equipped with a system of devices enabling the control of the size of air stream and its measurements. Due to the above-mentioned features, the installation enables the maintenance of repeatable values of the regulated factors. In order to measure the composition of the exhaust gases, samples were taken from a stub-pipe mounted in the chimney. The stub-pipe was located near the top of the combustion chamber. The concentrations of the analyzed gases were measured using the MGA 5 analyzer by MRU GmbH (Neckarsulm Germany). The analyzer measured CO_2, CO, NO, and SO_2 concentrations using infrared sensors. Additionally, in the case of NO_2, the analyzer was using a catalytic converter. The accuracy of the exhaust gas analyzer was ±5% of the measured value [41].

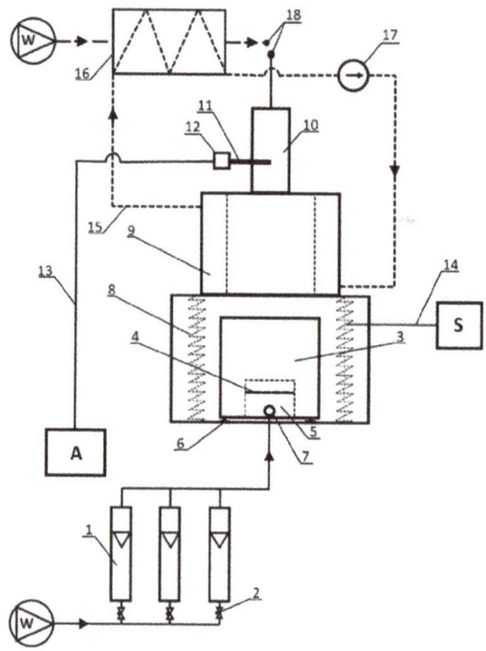

Figure 1. Test stand scheme: A—flue gas analyzer, S—boiler control system, W—fan, 1—rotameter, 2—valve, 3—moveable bed with grate, 4—grate, 5—surge tank (firepan), 6—rail, 7—air supply nozzle, 8—heating element (electrical), 9—water jacket, 10—discharge tunnel, 11—measurement probe, 12—probe head, 13—heated hose, 14—electric cables, 15—cooling water circulation, 16—water/air heat exchanger (cooler), 17—circulation pump, 18—surroundings.

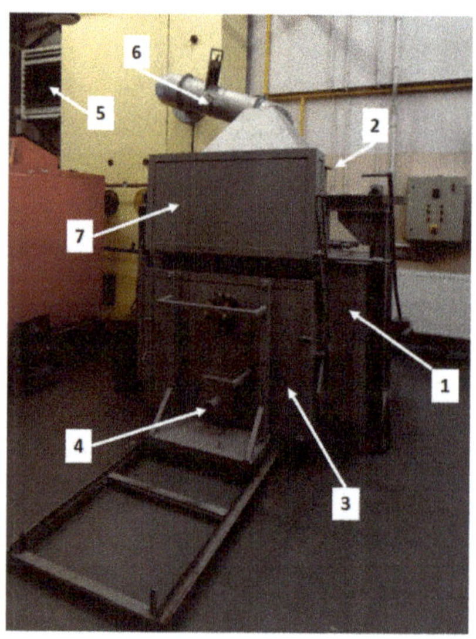

Figure 2. View of the test stand: 1—lower part of the stand (heated), 2—upper part of the stand (water-cooled), 3—movable bed with a grate (inserted), 4—connection point for the primary air duct, 5—water/air heat exchanger (cooler), 6—measurement point of exhaust gas composition, 7—raised furnace closure (lowered for the heating time of the stand).

2.3. Research Material

The research presented in this study concerns the co-firing of sludge with the composition given in Table 2 and hard coal with the composition shown in Table 3. The co-firing process was carried out in the furnace shown in Figure 1 (a fireplace with a fixed grate). The proportion of sludge in the burnt mixture was 15%, and the proportion of moisture in the sludge was 40%. The remaining parameters of the process were as follows: the initial thickness of the fuel layer on the grate was 10 cm, the theoretical ratio of excess air to combustion was 1.4, and the duration of the process was 40 min.

Table 2. Selected parameters of sludge used in the tests (d.w.—dry weight).

Specification	Unit	Values	Stand. Dev.	Determination Method/Standard
Calorific value d.w.	kJ/kg	10,330	680	PN-ISO 1928:2002 [42]
Combustible parts	% d.w.	62.04	0.61	PN-Z-15008-03:1993 [40]
		Elementary composition:		
Carbon	% d.w.	30.13	0.72	PN-Z-15008-05:1993 [43]
Hydrogen	% d.w.	4.35	0.34	PN-Z-15008-05:1993 [43]
Nitrogen	% d.w.	3.67	0.26	PN-G-04523:1992 [44]
Sulphur	% d.w.	1.41	0.16	PN-ISO 334:1997 [45]
Oxygen	% d.w.	19.30	0.85	Calculation method *

* oxygen content = 100%−carbon content %−hydrogen content %−nitrogen content %−sulfur content %−content of noncombustible parts %.

Table 3. Selected parameters of coal used in the research.

Specification	Unit	Values	Stand. Dev.	Determination Method/Standard
Calorific value d.w.	kJ/kg	26,790	2040	PN-ISO 1928:2002 [42]
Combustible parts	% d.w.	93.65	0.104	PN-Z-15008-03:1993 [40]
		Elementary composition:		
Carbon	%	73.01	0.97	PN-Z-15008-05:1993 [43]
Hydrogen	%	4.57	0.08	PN-Z-15008-05:1993 [43]
Nitrogen	%	1.53	0.04	PN-G-04523:1992 [44]
Sulphur	%	0.37	0.01	PN-ISO 334:1997 [45]
Oxygen	%	9.66	1.01	Calculation method *
Humidity	%	4.79	0.25	PN-G-04511:1980 [46]

* oxygen content = 100%−carbon content %−hydrogen content %−nitrogen content %−sulfur content %−content of non-combustible parts %−humidity content %.

2.4. Distribution of Primary Air

Under the above-mentioned conditions, the tests were carried out for three variants of air stream distribution, which involved the relative version $R \in [1/6, 5/12, 2/3]$, which in order to obtain absolute values should be multiplied by T.

In Figure 3, for the aforementioned R values, theoretical air stream distributions, determined according to the Equation (2), taking into account the dependencies (5) and (6), are presented. Figure 4 shows air streams in individual zones during the carried-out tests, which simulate the air flow distribution curves described by the Equation (2), for the R values taken into account. In each variant, the air was fed into 5 zones.

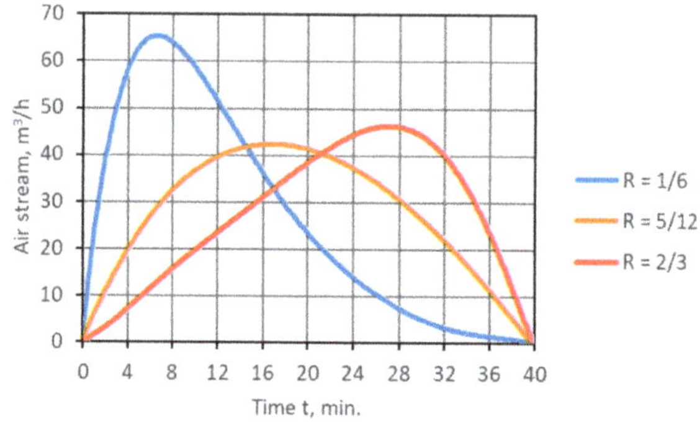

Figure 3. Theoretical distributions of air streams for the values R analyzed in the study.

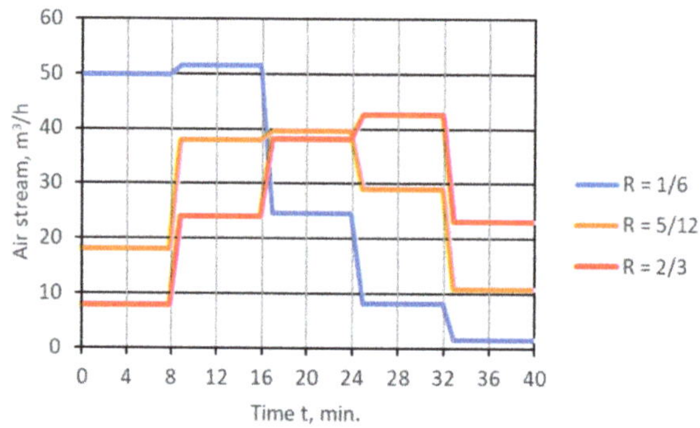

Figure 4. Average values of air streams delivered to individual zones during laboratory tests.

2.5. Sequence of Procedures in the Determination of Gaseous Products of the Combustion Process

- The following part of the article presents the results of tests carried out in compliance with the following procedure:
- Based on the composition and amount of the tested fuel and on the assumed ratio of excess air, the total amount of supplied air (V_{ad}) was calculated.
- For the assumed value of R, the air stream distribution in time $\dot{V}$ was determined in line withEquation (2) (Figure 3).
- Based on the Equation (10), the average values of air streams in individual zones $\dot{V}_{i,R}$ were calculated (Figure 4).
- The air streams, equal to the calculated values, were blown into the combustion chamber (they were controlled by means of rotameters).
- During the tests, the concentrations of the analyzed exhaust gas components were measured $e_{j,R,i,n}$; the measurements were made in subsequent time steps in individual zones (using an analyzer).

For the values of air streams and concentrations of exhaust gas components resulting from the measurement, from Equation (11), taking into account Equation (9), the emission streams of exhaust gas components were calculated.

3. Results

3.1. Emission Streams of Exhaust Gas Components

The following was assumed in the calculations: $I = 5$, $N = 8$, $\Delta t = 1$ min, and R $\in$ [1/6, 5/12, 2/3]. The obtained results are shown in Figures 5–8, respectively for CO_2, CO, NO_x, and SO_2. Each figure shows the emission streams corresponding to the three analyzed air distributions R. The information provided in the figures demonstrates the impact of air distribution both on the values of the emission streams of exhaust gas components obtained during the combustion of fuel and on the dynamics of the changes of these streams (i.e., their changes over time).

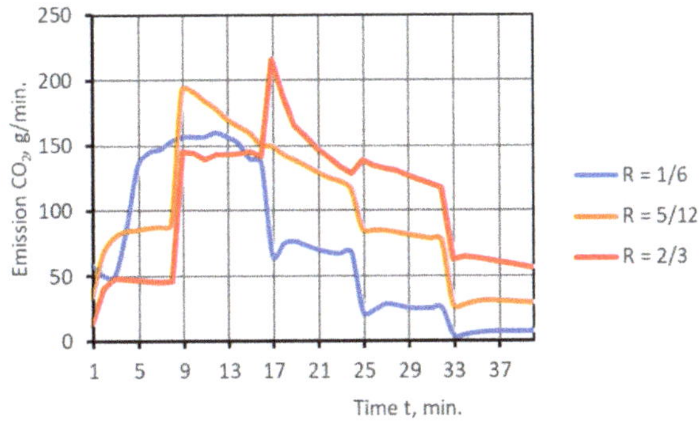

Figure 5. Dynamics of CO_2 emission determined during the tests at individual primary air distributions.

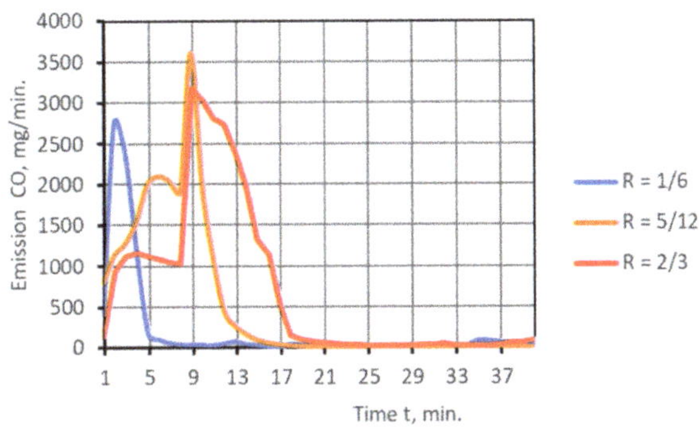

Figure 6. Dynamics of CO emission determined during the tests for individual primary air distributions.

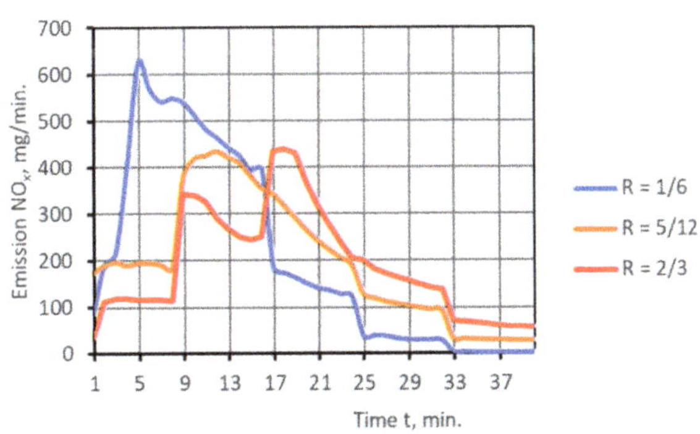

Figure 7. Dynamics of NO_x emission determined during the tests for individual primary air distributions.

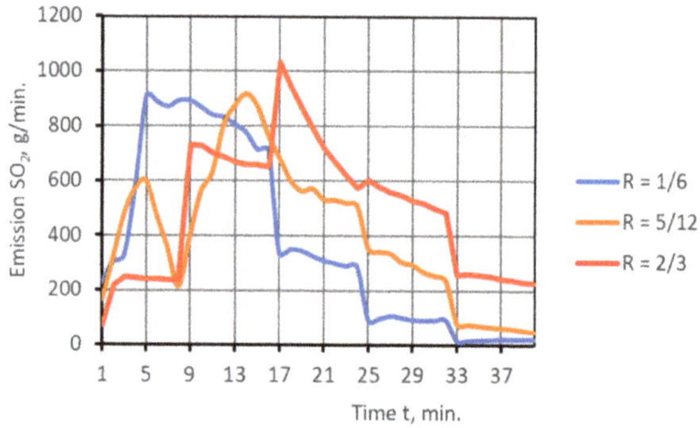

Figure 8. Dynamics of SO_2 emission determined during the tests for individual primary air distributions.

3.2. Total Emissions of Gaseous Combustion Products

The total emissions of the analyzed exhaust gas components were determined from the dependence (12) based on the above-mentioned measurement results and the adopted assumptions. The geometrical interpretation of the total emission of exhaust gas components is expressed by the fields under the graphs defining the emission stream of these components (see Figures 5–8). The obtained results are presented in Figure 9.

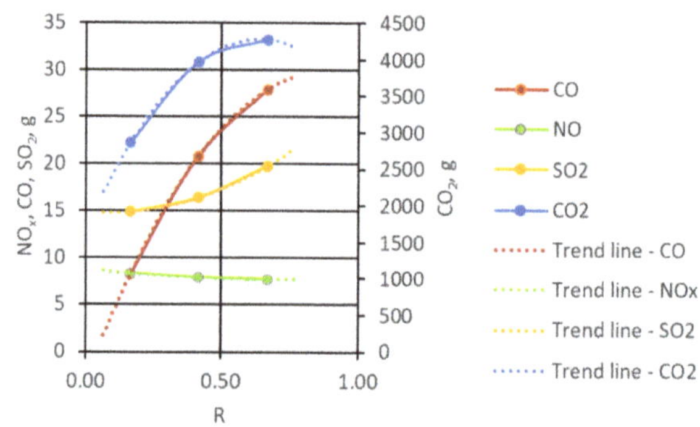

Figure 9. Curves describing changes in total emission of CO_2, CO, NO_x, and SO_2 as a function of the R parameter. Dashed lines—extrapolated trend lines beyond the test range (value 0.1 of the R parameter).

The dashed lines show the trend lines extrapolated beyond the test range (by the value of 0.1 of the parameter R).

3.3. Determination of the Share of Combustible Parts in the Slag

The contents of the combustible parts in the slag, measured for the three considered air distributions, are presented in Figure 10. The dashed line shows the trend line, which is the extrapolated trend line beyond the scope of the tests (by 0.1 of the R parameter).

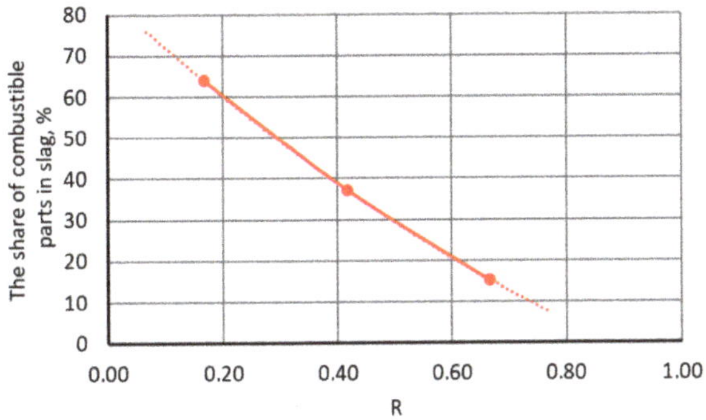

Figure 10. Curve describing changes in the content of combustible parts in the slag as a function of R parameter. The dashed line—extrapolated trend line beyond the test range (by the value of 0.1 of the R parameter).

4. Discussion

The following discussion is based on the results of the research carried out in a furnace chamber with a fixed grate, working at the operating parameters given in Section 2 of this paper.

The changes in air streams introduced during the tests caused disturbances in the combustion process, which contributed to the irregular course of the streams of gaseous emissions of exhaust gas components.

In the case of the curves showing the changes in CO_2 (Figure 5), the relationship between gas emissions and the changes in the primary air stream is clearly visible. And with the increase in parameter R, the occurrence of the largest CO_2 stream is increasingly ahead of the moment of maximum air stream supply.

Almost the entire emission of CO (Figure 6), regardless of the distribution of the supplied air, takes place up to the moment of the largest stream of supplied air (approx. to 8, 16 and 20 min, respectively). The combustion process with air separation, for which R = 1/6, is characterized by the shortest duration of high CO emissions and the lowest total emissions of CO_2 and CO. This may be the reason to conclude that the large air stream at the beginning of the combustion process did not allow the fuel to be properly ignited. For this reason, the amount of air quickly turned out to be sufficient to completely burn down the occurring CO to CO_2.

The largest streams of NO_x emission (Figure 7) are accompanied by a rapid decrease in the stream of CO emission. The above observation is consistent with the results of the research presented in [20,47,48].

The shapes of the curves showing changes in SO_2 emissions (Figure 8), in the case of combustion processes with air distribution for R = 1/6 and R = 2/3, approximately correspond to the shapes of the curves showing changes in CO_2 emissions (it is most evident in the case of air separation for R = 2/3). For the third combustion process, the discussed relationship can be observed only after 13 min.

The information provided in Figures 9 and 10 shows that within the range of the analyzed air distribution R from 1/6 to 2/3, as it increases, a rise in the emissions of CO_2, CO and SO_2 was reported, respectively by 53%, 125%, and 27%, as well as a drop of NO_x emission and the share of combustible parts in the slag by 12% and 79%, respectively. This allows us to assess the impact of air distribution on incomplete and fragmentary fuel combustion and to evaluate the ecological harmfulness of the conducted combustion processes.

The following general comments and observations result from the conducted analyses:

- The generalized air distribution curve corresponds well with the practical methods of air supply (distribution) in industrial grate furnaces.

- The time or place of supplying the maximum air stream and its total demand are sufficient parameters to determine the shape of the air distribution function.
- Air distribution has a significant impact on the quantity and quality of gaseous and solid products of the combustion process.
- By applying the proper air distribution, it is possible to significantly reduce the emission of selected gaseous substances and to increase the energy efficiency of the furnace operation.
- In order to determine the most advantageous location of the maximum stream, the furnace characteristics should be determined each time for a given fuel (e.g., coal with specific properties).

When analyzing the obtained measurement results, one should notice the presence of too-high contents of combustible parts in the slag and the related too-low values of CO_2 emission. It is caused by too short fuel combustion assumed in the tests. The resulting fact made it possible to more clearly demonstrate the influence of the method of air supply in the grate furnace on the emission of solid and gaseous fuel combustion products.

Author Contributions: Conceptualization, M.K. and J.K.; methodology, M.K. and J.K.; validation, M.K.; formal analysis, J.K.; investigation, M.K.; resources, M.K.; data curation, M.K. and J.K.; writing—original draft preparation, M.K. and J.K.; writing—review and editing, M.K. and J.K.; visualization, M.K.; supervision, J.K.; project administration, M.K.; funding acquisition, M.K. All authors have read and agreed to the published version of the manuscript.

Funding: The APC was funded by the Silesian University of Technology from the research subsidy (SUBB) for 2023, number 08/030/BK_23/0116.

Data Availability Statement: Not applicable.

Conflicts of Interest: The authors declare no conflict of interest.

References

1. Madanayake, B.N.; Gan, S.; Eastwick, C.; Ng, H.K. Biomass as an energy source in coal co-firing and its feasibility enhancement via pre-treatment techniques. *Fuel Process. Technol.* **2017**, *159*, 287–305. [CrossRef]
2. Ozawa, M.; Asano, H. (Eds.) *Advances in Power Boilers*; Elsevier: Amsterdam, The Netherlands, 2021. [CrossRef]
3. Orłowski, P.; Dobrzański, W.; Szwarc, E. *Kotły Parowe, Konstrukcja i Obliczenia*; Wydawnictwa Naukowo-Techniczne: Warszawa, Poland, 1979.
4. Fang, J.; Li, G.; Aunan, K.; Vennemo, H.; Seip, H.M.; Oye, K.A.; Beér, J.M. A proposed industrial-boiler efficiency program in Shanxi: Potential CO2-mitigation, health benefits and associated costs. *Appl. Energy* **2002**, *71*, 275–285. [CrossRef]
5. Fang, J.; Zeng, T.; Yang, L.I.; Oye, K.A.; Sarofim, A.F.; Beér, J.M. Coal utilization in industrial boilers in China —A prospect for mitigating CO2 emissions. *Appl. Energy* **1999**, *63*, 35–52. [CrossRef]
6. Lin, P.; Ji, J.; Luo, Y.; Wang, Y. A non-isothermal integrated model of coal-fired traveling grate boilers. *Appl. Therm. Eng.* **2009**, *29*, 3224–3234. [CrossRef]
7. Greinert, A.; Mrówczyńska, M.; Grech, R.; Szefner, W. The Use of Plant Biomass Pellets for Energy Production by Combustion in Dedicated Furnaces. *Energies* **2020**, *13*, 463. [CrossRef]
8. Variny, M.; Varga, A.; Rimár, M.; Janošovský, J.; Kizek, J.; Lukáč, L.; Jablonský, G.; Mierka, O. Advances in Biomass Co-Combustion with Fossil Fuels in the European Context: A Review. *Processes* **2021**, *9*, 100. [CrossRef]
9. Zhang, X.; Chen, Q.; Bradford, R.; Sharifi, V.; Swithenbank, J. Experimental investigation and mathematical modelling of wood combustion in a moving grate boiler. *Fuel Process. Technol.* **2010**, *91*, 1491–1499. [CrossRef]
10. Yin, C.; Rosendahl, L.A.; Kær, S.K. Grate-firing of biomass for heat and power production. *Prog. Energy Combust. Sci.* **2008**, *34*, 725–754. [CrossRef]
11. Pham, T.P.T.; Kaushik, R.; Parshetti, G.K.; Mahmood, R.; Balasubramanian, R. Food waste-to-energy conversion technologies: Current status and future directions. *Waste Manag.* **2015**, *38*, 399–408. [CrossRef]
12. Jančauskas, A.; Buinevičius, K. Combination of Primary Measures on Flue Gas Emissions in Grate-Firing Biofuel Boiler. *Energies* **2021**, *14*, 793. [CrossRef]
13. Regueiro, A.; Patiño, D.; Porteiro, J.; Granada, E.; Míguez, J.L. Effect of Air Staging Ratios on the Burning Rate and Emissions in an Underfeed Fixed-Bed Biomass Combustor. *Energies* **2016**, *9*, 940. [CrossRef]
14. Kozioł, M. Możliwości Substytucji Węgla Paliwami z Odpadów w Warunkach Polskich. *Rynek Energii* **2022**, *3*, 64–70. Available online: http://rynek-energii.pl/pl/RE160 (accessed on 22 September 2022).
15. Makarichi, L.; Jutidamrongphan, W.; Techato, K.-A. The evolution of waste-to-energy incineration: A review. *Renew. Sustain. Energy Rev.* **2018**, *91*, 812–821. [CrossRef]

16. Song, J.; Sun, Y.; Jin, L. PESTEL analysis of the development of the waste-to-energy incineration industry in China. *Renew. Sustain. Energy Rev.* **2017**, *80*, 276–289. [CrossRef]
17. Leckner, B.; Lind, F. Combustion of municipal solid waste in fluidized bed or on grate—A comparison. *Waste Manag.* **2020**, *109*, 94–108. [CrossRef] [PubMed]
18. Dai, J.; Sokhansanj, S.; Grace, J.R.; Bi, X.; Lim, C.J.; Melin, S. Overview and some issues related to co-firing biomass and coal. *Can. J. Chem. Eng.* **2008**, *86*, 367–386. [CrossRef]
19. Xin-Gang, Z.; Gui-Wu, J.; Ang, L.; Yun, L. Technology, cost, a performance of waste-to-energy incineration industry in China. *Renew. Sustain. Energy Rev.* **2016**, *55*, 115–130. [CrossRef]
20. Wilk, R.K. *Low-Emission Combustion*; Wydawnictwo Politechniki Śląskiej: Gliwice, Poland, 2002.
21. Nugraha, M.; Saptoadi, H.; Hidayat, M.; Andersson, B.; Andersson, R. Particulate Matter Reduction in Residual Biomass Combustion. *Energies* **2021**, *14*, 3341. [CrossRef]
22. Vakkilainen, E.K. *Steam Generation from Biomass. Construction and Design of Large Boilers*; Butterworth-Heinemann/Elsevier: Oxford, UK, 2016. [CrossRef]
23. Hoang, Q.N.; Vanierschot, M.; Blondeau, J.; Croymans, T.; Pittoors, R.; Van Caneghem, J. Review of numerical studies on thermal treatment of municipal solid waste in packed bed combustion. *Fuel Commun.* **2021**, *7*, 100013. [CrossRef]
24. Koziol, J.; Kozioł, M. Determining operating characteristics of co-firing processes in grate furnaces. *Fuel* **2019**, *258*, 116164. [CrossRef]
25. Jiao, L.; Kuang, M.; Chen, Y.; Liu, S.; Wang, X. Detailed measurements of in-furnace gas temperature and species concentration distribution regarding the primary-air distribution mode in a spreader and reversal chain-grate furnace. *Energy* **2021**, *235*, 121384. [CrossRef]
26. Hoang, Q.N.; Van Caneghem, J.; Croymans, T.; Pittoors, R.; Vanierschot, M. A novel comprehensive CFD-based model for municipal solid waste incinerators based on the porous medium approach. *Fuel* **2022**, *326*, 124963. [CrossRef]
27. Zhuang, J.; Tang, J.; Aljerf, L. Comprehensive review on mechanism analysis and numerical simulation of municipal solid waste incineration process based on mechanical grate. *Fuel* **2022**, *320*, 123826. [CrossRef]
28. Frey, H.-H.; Peters, B.; Hunsinger, H.; Vehlow, J. Characterization of municipal solid waste combustion in a grate furnace. *Waste Manag.* **2003**, *23*, 689–701. [CrossRef] [PubMed]
29. Scholz, R.; Beckmann, M.; Schulenburg, F. *Entwicklungsmöglichkeiten der Prozeßführung bei Rostsystemen zur Abfallbehandlung*; FDBR-Symposium: Rostock, Germany, 1996.
30. Wissing, F.; Wirtz, S.; Scherer, V. Simulating municipal solid waste incineration with a DEM/CFD method—Influences of waste properties, grate and furnace design. *Fuel* **2017**, *206*, 638–656. [CrossRef]
31. Gu, T.; Ma, W.; Guo, Z.; Berning, T.; Yin, C. Stable and clean co-combustion of municipal sewage sludge with solid wastes in a grate boiler: A modeling-based feasibility study. *Fuel* **2022**, *328*, 125237. [CrossRef]
32. Costa, M.; Massarotti, N.; Mauro, A.; Arpino, F.; Rocco, V. CFD modelling of a RDF incineration plant. *Appl. Therm. Eng.* **2016**, *101*, 710–719. [CrossRef]
33. Tu, Y.; Zhou, A.; Xu, M.; Yang, W.; Siah, K.B.; Prabakaran, S. Experimental and numerical study on the combustion of a 32 MW wood-chip grate boiler with internal flue gas recirculation technology. *Energy Procedia* **2017**, *143*, 591–598. [CrossRef]
34. Rajh, B.; Yin, C.; Samec, N.; Hriberšek, M.; Zadravec, M. Advanced modelling and testing of a 13 MW th waste wood-fired grate boiler with recycled flue gas. *Energy Convers. Manag.* **2016**, *125*, 230–241. [CrossRef]
35. Xia, Z.; Shan, P.; Chen, C.; Du, H.; Huang, J.; Bai, L. A two-fluid model simulation of an industrial moving grate waste incinerator. *Waste Manag.* **2020**, *104*, 183–191. [CrossRef]
36. Karim, R.; Naser, J. CFD modelling of combustion and associated emission of wet woody biomass in a 4 MW moving grate boiler. *Fuel* **2018**, *222*, 656–674. [CrossRef]
37. Jiang, M.; Lai, A.; Law, A. Solid Waste Incineration Modelling for Advanced Moving Grate Incinerators. *Sustainability* **2020**, *12*, 8007. [CrossRef]
38. Zhou, A.; Xu, H.; Tu, Y.; Zhao, F.; Zheng, Z.; Yang, W. Numerical investigation of the effect of air supply and oxygen enrichment on the biomass combustion in the grate boiler. *Appl. Therm. Eng.* **2019**, *156*, 550–561. [CrossRef]
39. Waldner, M.; Halter, R.; Sigg, A.; Brosch, B.; Gehrmann, H.; Keunecke, M. Energy from Waste—Clean, efficient, renewable: Transitions in combustion efficiency and NOx control. *Waste Manag.* **2013**, *33*, 317–326. [CrossRef] [PubMed]
40. PKN. PN-Z-15008-03:1993 Badania Właściwości Paliwowych—Oznaczanie Zawartości Części Palnych i Niepalnych. Available online: https://sklep.pkn.pl/pn-z-15008-03-1993p.html (accessed on 14 July 2021).
41. *MRU Sp. z o.o.: Instrukcja Obsługi i Serwisu MGA 5 (MGA 5 Flue Gas Analyzer User Manual)*; MRU GmbH: Neckarsulm, Germany, 2006.
42. PKN: PN-ISO 1928:2002 Paliwa Stałe—Oznaczanie Ciepła Spalania Metodą Spalania w Bombie Kalorymetrycznej i Obliczanie Wartości Opałowej. Available online: https://sklep.pkn.pl/pn-iso-1928-2002p.html?options=cart (accessed on 20 July 2021).
43. PKN: PN-Z-15008-05:1993 Odpady Komunalne Stałe—Badania Właściwości Paliwowych—Oznaczanie Zawartości Węgla i Wodoru. Available online: https://sklep.pkn.pl/pn-z-15008-05-1993p.html (accessed on 14 July 2021).
44. PKN: PN-G-04523:1992 Paliwa Stałe—Oznaczanie Zawartości Azotu Metodą Kjeldahla. Available online: https://sklep.pkn.pl/pn-g-04523-1992p.html?options=cart (accessed on 14 July 2021).

45. PKN: PN-ISO 334:1997 Paliwa Stałe—Oznaczanie Siarki Całkowitej—Metoda Eschki. Available online: https://sklep.pkn.pl/pn-iso-334-1997p.html?options=cart (accessed on 14 July 2021).
46. PKN: PN-G-04511:1980 Paliwa Stałe—Oznaczanie Zawartości Wilgoci. Available online: https://sklep.pkn.pl/pn-g-04511-1980p.html (accessed on 5 July 2021).
47. Kruczek, H.; Modliński, Z.; Ferens, W. Study of NOx formation for lean and rich stirred reactors flames in the presence of sulphur species. In Proceedings of the XIIIth International Symposium on Combustion Processes, Kraków, Poland, 22–24 September 1993.
48. Scholz, R.; Beckmann, M.; Davidivic, M.; Busch, M.; Schmidt, W. *Vergasung von Restabfall auf Rosten*; Heft 13; Veröffentlichungen des Zentrums für Abfallforschung der Technischen Universität Braunschweig: Braunschweig, Germany, 1998.

energies

Article

Combustion of Lean Methane/Propane Mixtures with an Active Prechamber Engine in Terms of Various Fuel Distribution

Ireneusz Pielecha * and Filip Szwajca

Faculty of Civil and Transport Engineering, Poznan University of Technology, 60-965 Poznan, Poland; filip.szwajca@put.poznan.pl
* Correspondence: ireneusz.pielecha@put.poznan.pl; Tel.: +48-61-224-4502

Abstract: The possibilities for reducing the fuel consumption of internal combustion engines focus mainly on developing combustion systems, as one such solution is a two-stage combustion system using jet ignition. The combustion of gaseous mixtures with a high excess air ratio leads to an increase in overall efficiency and a reduction in the emissions of selected exhaust components. In such a convention, gas combustion studies were conducted in the methane/propane configuration. Using an active prechamber where spark plugs were placed and direct injection through a check valve, the fuel dose was minimized into the prechamber. The tests were conducted for a constant center of combustion (CoC). The combustion process in both the prechamber and main chamber was analyzed using a test stand equipped with a 0.5 dm^3 single-cylinder engine. The engine was controlled by varying the fuel supply to the prechamber and main chamber in excess air ratio λ = 1.3–1.8. The study analyzed thermodynamic indices such as the combustion pressure in both chambers, based on which the SoC in both chambers, the rate and amount of heat released, AI05, AI90 and, consequently, the indicated efficiency were determined. Based on the results, it was found that the use of CH_4/C_3H_8 combination degraded the thermodynamic indicators of combustion more than using only the base gas (methane). In addition, the stability of the engine's operation was decreased. The advantage of using propane for the prechamber is to obtain more beneficial ecological indicators. For the single-fuel system, a maximum indicated efficiency of more than 40% was obtained, while with the use of propane for the prechamber, a maximum of 39.3% was achieved.

Keywords: TJI combustion; prechamber; methane/propane combustion

Citation: Pielecha, I.; Szwajca, F. Combustion of Lean Methane/Propane Mixtures with an Active Prechamber Engine in Terms of Various Fuel Distribution. *Energies* **2023**, *16*, 3608. https://doi.org/10.3390/en16083608

Academic Editors: Tomasz Czakiert and Monika Kosowska-Golachowska

Received: 13 March 2023
Revised: 18 April 2023
Accepted: 19 April 2023
Published: 21 April 2023

1. Introduction

The development of internal combustion engines focused on reducing fuel consumption is primarily leading to improvements in lean-burn technology. Gaseous fuels are mainly becoming attractive as an energy factor due to their lower carbon content, effectively reducing carbon dioxide and particulate emissions into the atmosphere [1].

The combustion of methane in spark-ignition engines in stoichiometric mode is a relatively common solution applied in many types of propulsion systems. An expanded opportunity to improve engine ecology and reduce fuel consumption is provided using a two-stage combustion system—TJI [2–5]. This system can be classified into passive and active prechambers [6]. The passive prechamber is filled with homogeneous fuel–air mixtures from the main chamber during the compression stroke. The active prechamber system is integrated with an auxiliary fuel-metering device to accurately control the equivalence ratio of the stratified mixture. Thus, the passive prechamber and active prechamber systems are also named homogeneous prechamber and stratified prechamber systems, respectively [6]. The above shows much greater benefits when using an active prechamber system. Ignition mechanisms [7–9] and inter-chamber flows have been well described and explained.

Combustion with turbulent jet ignition systems is carried out over a wide range of excess air ratios. In the range of excess air ratio 1–1.5, the authors of the article [10]

investigated the combustion of various fuels in a prechamber using a single-cylinder test engine. Additionally, in the range of excess air ratio from 1.5 to 1.9, the effect of the cross-sectional area of the prechamber discharge holes was studied. The tests were conducted on an engine fueled with natural gas [11]. Studies on the flammability of the air–fuel mixture using a constant-volume chamber and the TJI system were conducted. In addition to base natural gas, hydrogen was also burned [12]. On another model test stand, specifically a rapid compression machine, different configurations of the ignition system, such as PC spark plug, TJI, and a three-stage system, were compared for an excess air ratio of 1.5 [13]. In another study [14], Nanosecond Repetitively Pulsed Discharge (NRPD) ignition systems and turbulent jet ignition (TJI) were tested in a constant volume chamber with an excess air ratio of up to 1.8.

The effects of co-combustion of methane and propane are not well recognized, as evidenced by the small number of published research results. Studies of these fuels as an additive to diesel fuel have been conducted. Their use reduced fuel consumption by 21% or 15%, respectively [15]. The same reduction in NO_x emissions (by 56–57%) was obtained regardless of the additive used.

The much higher density of propane than methane (1.964 versus 0.715 kg/m^3) and a boiling point of −42 °C (methane −162 °C) cause difficulties in getting propane into the cylinder in gaseous form [16]. It is necessary to keep appropriate conditions such as low pressure.

One of the concepts for burning lean mixtures is the so-called hybrid combustion [17]. It uses a micro-flame-ignited (MFI) in dimethyl ether (DME) direct injection system and gasoline at a lambda excess air ratio of 2.0. Li at el. [17] conducted studies of DME combustion with the hybrid combustion of DME and gasoline. As a result, the indicated mean effective pressure (IMEP) and cyclic variation are reduced in the double direct injection conditions.

Modeling of the methane combustion process with the TJI system was conducted by Distaso et al. [18]. The research was carried out using an active prechamber at $\lambda = 1.3$. Such a value was found to be the limiting value in a standard engine. The prechamber (cylindrical in shape) was placed angular to the cylinder axis. As a result of the conducted exhaust emission analyses, it was determined that in both chambers at the exhaust valve open EVO, the mass fraction of CO_2 produced is almost the same. The production of CO and HC in the prechamber is significantly higher (by two orders of magnitude). The mass fraction share of NO_x in the prechamber is an order of magnitude smaller than in the main chamber.

A future direction for fueling internal combustion engines may be using ammonia. Liu et al. [19] report that using a TJI system for ammonia combustion improves the stability of engine operation and makes it possible to obtain higher IMEP values with respect to a reference engine. However, Vinod et al. list many barriers to the development of this fuel, such as long ignition delay, low flame development rate, and low reactivity [20]. The quality of the combustion process can be improved by adding methane to ammonia [21] or by using a reactivity controlled turbulent jet ignition (RCTJI) system [22]. Research by Zhang et al. [21] indicates that a 10 to 20% methane addition value leads to an increase in combustion pressure and an increase in the average rate of pressure rise.

2. Aim and Scope of the Study

Co-combustion of gaseous fuels in a two-stage combustion system is a relatively unrecognized solution. For this reason, the authors decided to conduct a comparative experimental study using a single-cylinder research engine. The TJI system used a single-fuel operation where the fuel was methane, then the strategy was changed to a dual-fuel operation, and propane was supplied to the ignition chamber instead of methane. So, the comparison subject is the effect of changing the fuel supplied to the ignition chamber. In the research program on the effect of fuel type, large changes in the dose of fuel supplied to the prechamber with varying values of the excess air ratio were realized.

The questions sought to be answered regarding the thermodynamics of the process were (1) how the excess air ratio affects the combustion process, and (2) how the dose of fuel fed into the prechamber affects combustion efficiency.

3. Materials and Methods

3.1. Test Stand

A single-cylinder AVL 5804 research engine with Eddy current engine dynamometers was used to study the combustion process. The engine has an ignition system that allows real-time control of ignition advance angle and coil charging time. The cylinder head of the engine was adapted to a two-stage combustion system (Figure 1). The adaptation involved expanding the bore to fit the prechamber along with the direct fuel delivery system and pressure sensor. An active prechamber system with fuel delivery through a check valve was used. A prechamber with a 1.7 mm diameter straight 6-hole and an M10 spark plug was employed. Prechamber volume is 5.93% of clearance volume V_c. Due to the different properties of the fuels, methane was supplied to the main chamber at a pressure of 9 bar regardless of the engine mode. In single-fuel mode, methane was supplied to the prechamber at a pressure of 3 bar, while in dual-fuel mode the propane pressure was 1 bar. Fuel injection simultaneously occurs into the intake manifold and the prechamber during the intake stroke 300 °CA before TDC. An injector located in the intake manifold delivers fuel to the main chamber (this chamber is determined by the volume of the cylinder after the intake valves are closed)—Figure 1. Under these conditions, the use of higher injection pressure is not necessary. Other engine specifications are shown in Table 1 and Figure 2.

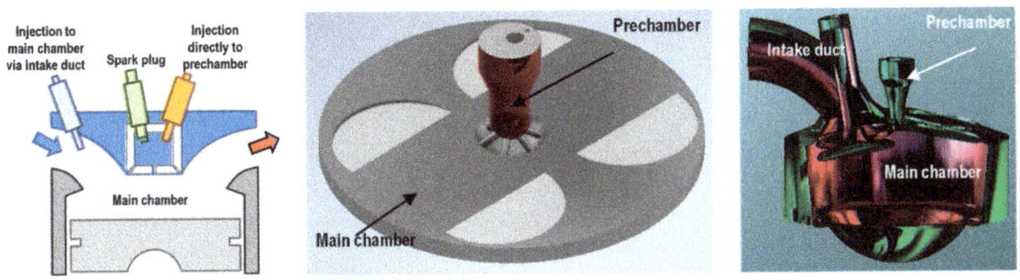

(a) **(b)** **(c)**

Figure 1. Scheme of cylinder head: (**a**) model of active prechamber; (**b**) view of main and prechamber; (**c**) view of intake duct and combustion chamber (main and prechamber).

Table 1. Technical data of the single-cylinder test engine used.

Parameter	Unit	Value
Engine	–	1-cyl., 4-valve, SI, TJI
Displacement	dm^3	0.5107
Bore × stroke	mm	85 × 90
Compression ratio	–	14.5
Fueling	–	Prechamber: EM injector Main chamber EM injector
Prechamber	–	2.35 cm^3 (5.93% of V_c)
Air system	–	Naturally aspirated

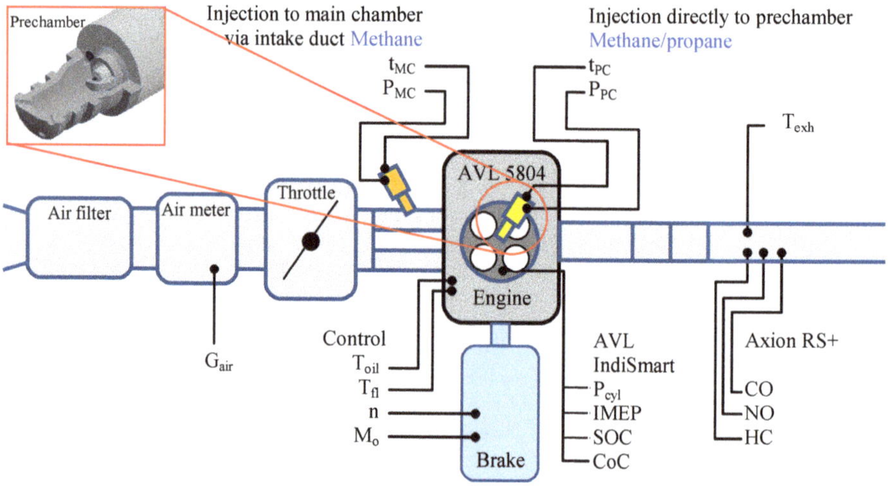

Figure 2. Test stand layout with a 2-stage combustion system fueled by methane and propane.

1. Variable control parameters:
    - Excess air ratio (λ = 1.3; 1.5; 1.8);
    - Dose of fuel to prechamber: energy value of fuel for prechamber: 10; 20; 30; 40; 45; 50; and 60 J (while keeping the total energy value of fuel supplied to the engine constant).
2. Constant control parameters:
    - Engine speed: n = 1500 rpm;
    - Total fuel dose: qo = 13.5 mg (energy = 675 J).

In view of the different calorific values of the fuels (methane and propane—Table 2), the energy content of the fuel dose was determined, rather than the mass directly.

Table 2. Properties of methane and propane [23,24].

Property	Methane	Propane
Chemical formula	CH_4	C_3H_8
Lower flammable limit [%]	5	2.1
Upper flammable limit [%]	15.4	9.5
Flammable range [%]	10.4	7.4
Risk index of explosion [−]	2.0	3.524
Minimum ignition energy [mJ]	0.21–0.30	0.25–0.27
Auto ignition temperature [°C]	580	480
Stoichiometric air/fuel ratio [−]	17.19	15.67
Adiabatic flame temperature [°C]	1963	1980
Laminar burning velocity [cm/s]	37	39–43
Lower heating value [MJ/kg]	50.0	46.35

Studies of thermodynamic parameters of combustion were conducted using combustion pressure sensors for the main chamber, AVL GH14D (0–250 bar), and in the prechamber, Kistler 6081 (0–250 bar), whose signals were recorded using AVL IndiSmart (8-channel + IFEM amplifiers) together with AVL crank angle (364C01; 0.1 deg). Gas flow rates: air (Sensycon Sensyflow P; 0–400 kg/h; error < ±0.8%); combustible gases into the prechamber (Bronkhorst 111B; 0.1–100 g/h; accuracy ±0.5% RD plus ±0.1% FS) and into the main chamber (Micro Motion ELITE CMFS010M; 0.1–2 kg/h; accuracy ±0.25%). Gas feed settings were adjusted using a system for controlling the timing and start of injection (Mechatronics Control Gas Injectors). Exhaust gas analysis was carried out using an Axion RS+

analyzer (CO: 0–10% accuracy ±0.02% abs, HC: 0–4000 ppm accuracy ±4 ppm abs, NO: 0–4000 ppm accuracy ±5 ppm abs), a typical portable emissions measurement system (PEMS—Portable Emissions Measurement System) from Global MRV. Exhaust gases were measured using the following methods: CO and HC—spectrometric via analyzer (NDIR) and NO—electrochemical.

3.2. Method of Analyzing Research Results

Methane was chosen as the primary fuel supplied to the cylinder due to its lower carbon content in the molecule. The research work included feeding methane or propane into the prechamber.

By changing the value of the excess air ratio and the type of fuel, engine operating conditions change even within the range of the same test point. Controlling engine operation by keeping the ignition angle constant, the maximum pressure angle constant, or the combustion center constant is possible. The last indicator, defined as the angle at which 50% of the heat is released (its value was set at 8 deg aTDC), was chosen:

$$\mathrm{CoC} = \alpha \text{ at } 0.5 \times \int_{\mathrm{SOC}}^{\mathrm{EOC}} \frac{dQ_{net}}{d\alpha} d\alpha, \quad (1)$$

where SOC—start of combustion; EOC—end of combustion. In a similar way, the beginning of combustion (the angle at which 5% of the heat is exerted) and the end of combustion (the angle at which 90% of the heat is exerted) were determined.

A criterion for the stability of engine operation has also been defined as the unevenness of operation determined by the value of the coefficient of variation CoV(IMEP) < 3.0% [25]. Older sources give this indicator a value of 10% [26] or this value is given in a range [27]. This indicator was defined as

$$\mathrm{CoV(IMEP)} = 100 \times \frac{\sigma(\mathrm{IMEP})}{\mu(\mathrm{IMEP})}, \quad (2)$$

where σ and μ are the standard deviation and the mean value, respectively, over a number of consecutive combustion cycles (analysis applies to 100 consecutive cycles).

Other thermodynamic indicators were determined as follows:

1. Heat release rate $\left(\frac{dQ_{net}(\alpha)}{d\alpha}\right)$

$$\frac{dQ_{net}(\alpha)}{d\alpha} = \frac{\gamma}{\gamma-1} P(\alpha) \frac{dV(\alpha)}{d\alpha} + \frac{1}{\gamma-1} V(\alpha) \frac{dP(\alpha)}{d\alpha}, \quad (3)$$

where P is the instantaneous cylinder pressure, α is the crank angle, γ is the ratio of the specific heats, and V is the instantaneous cylinder volume.

2. Indicative power (N_i):

$$N_i = \frac{V_s \times \mathrm{IMEP} \times n}{\tau}, \quad (4)$$

where V_s—engine displacement, n—engine speed, and τ—cyclicality of engine operation.

3. Specific fuel consumption (g_i):

$$g_i = \frac{G_{MC} + G_{PC}}{N_i}, \quad (5)$$

where G—fuel consumption in the main chamber (MC) and prechamber (PC), respectively.

4. Indicative efficiency (η_i):

$$\eta_i = \frac{1}{g_i \times L_{HV}}, \quad (6)$$

where L_{HV}—heating value of methane.

5. Specific emissions of exhaust components (e_i):

$$e_i = \frac{a_i \times C_i \times (G_a + G_{MC} + G_{PC})}{N_i}, \quad (7)$$

where i = CO, THC, NO, C_i—concentration; CO, THC, NO, G_a—air consumption; G_{MC}, G_{PC}—fuel consumption in both chambers; a_i—density ratios (CO = 0.000966; C_{THC} = 0.000479, C_{NO} = 0.001587) [28].

An illustrative curve of the measured quantities recorded during the tests is shown in Figure 3. In addition to the cylinder and prechamber pressures, the duration of the pulse controlling the injectors and ignition coil is also shown. Fuel was injected into both combustion chambers at different pressures. As can be seen from the figure, the dose injection time into the prechamber is significantly shorter than the fuel injection time into the main chamber.

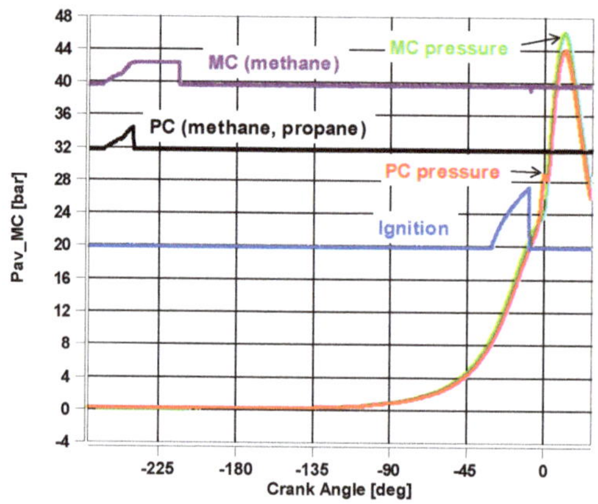

Figure 3. An example of the waveform of the recorded signals with the description of the fuels fed to both combustion chambers (start of fuel feeding to both chambers at the angle α = 260 deg bTDC).

4. Thermodynamic Analysis of System Operation Fueled by Different Fuels

4.1. Cylinder Pressure

According to the conducted tests, the in-cylinder pressure curves were obtained as an average of the 100 recorded cycles. Due to the fact that methane was always injected into the main chamber, Figure 4 only shows what type of fuel was injected into the prechamber. The blue color represents the case of fueling the pre-combustion chamber with methane, while the red color represents the case of fueling it with propane. The diagrams also include information on the share of energy contained in the fuel delivered to the prechamber. With the excess air ratio with the smallest analyzed value (λ = 1.3), feeding methane into the prechamber resulted in slightly higher values of maximum pressure in the cylinder (in the main chamber). At λ = 1.5, the maximum pressure values are very similar. Operation of the engine with an excess air ratio of λ = 1.8 results in the smallest dose of methane fed into the prechamber not igniting the fuel in the cylinder (no line in Figure 4). Only at a dose of qo_PC = 30 J and λ = 1.8 is the value of the maximum pressure during propane combustion in the PC greater than that of methane combustion.

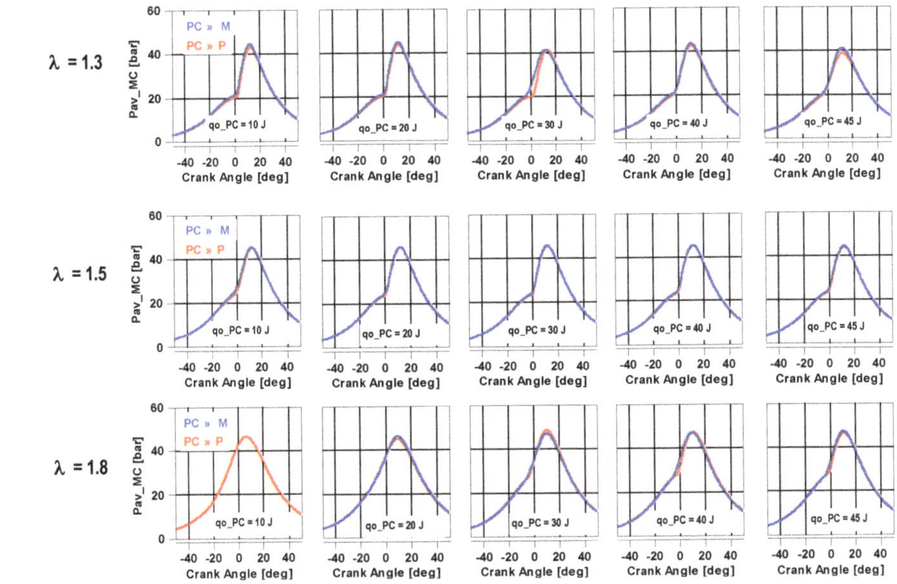

Figure 4. In-cylinder (main chamber) pressure curves Pav_MC averaged from 100 cycles of combustion pressure in the cylinder when the prechamber was fueled with methane or propane (methane was always supplied to the main chamber via injector placed in intake duct).

4.2. Analysis of Engine Operation Stability

For all engine operating points, the coefficient of variation of the indicative mean effective pressure CoV(IMEP) was determined according to Equation (2). The data presented (Figure 5) show that the largest values of the engine's operating irregularity are related to a large dose of fuel fed into the prechamber at λ = 1.3, independent of the fuel type in the prechamber. As mentioned earlier, the misfire of the fuel in the cylinder, shown by the large CoV(IMEP) value, also occurs at λ = 1.8 and a low fuel dose. At this point, CoV(IMEP) = 12% was obtained during propane combustion, while no combustion occurred during methane combustion (in the prechamber), which is why there is no point in Figure 5b.

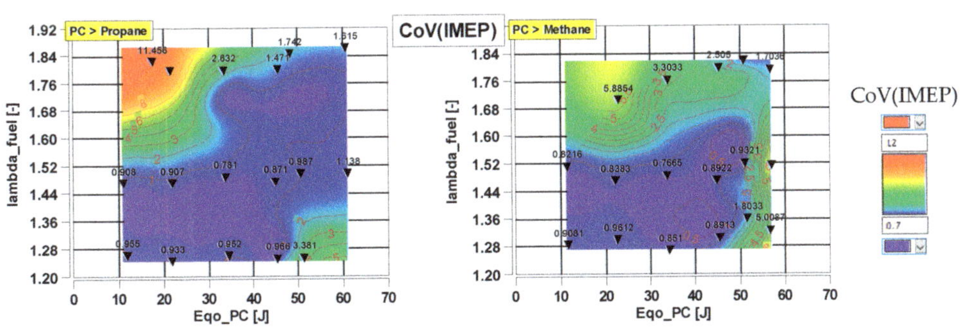

(a) (b)

Figure 5. Interpolated maps showing the stability of engine operation represented by the coefficient of variation CoV(IMEP): (**a**) when burning propane in the prechamber; (**b**) when burning methane in the prechamber.

As shown in Figure 5, the most stable engine operation is at λ = 1.5 over the entire range of changes in the fuel dose to the prechamber qo_PC. The greatest unstable operation is observed with a large value of λ = 1.8 and a low dose of propane to PC (Figure 5a). A large excess air ratio and a low fuel dose do not promote the ignitability of the charge. The same is true for methane combustion—with the above conditions (λ = 1.8, qo_PC = min) there is no ignitability of the main charge.

Therefore, the next figure shows the conditions that are the most (Figure 6a) and least favorable (Figure 6b) for ignition of the charge. The most favorable conditions for combustion (most stable engine operation) occur at the average value of the fuel dose to the PC (qo = 30 J). The value of CoV(IMEP) is below 0.8%. The least favorable conditions result in CoV(IMEP) values well above 5% and above 9%. It can be considered that at CoV(IMEP) = 9.3%, the peak pressure differences are almost 100%. At CoV(IMEP) = 5.8%, the maximum variations are 26 bar, or about 75%.

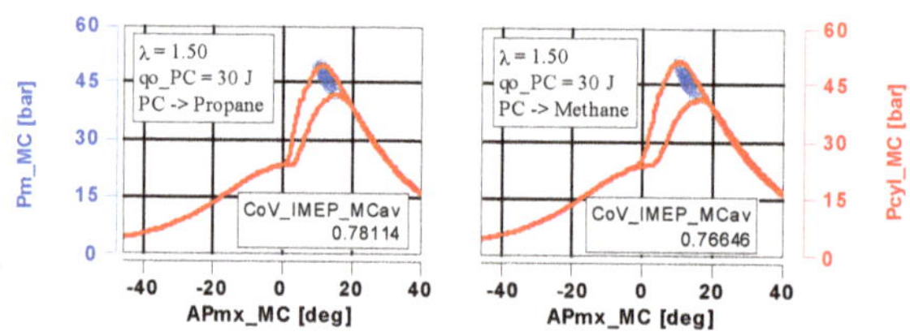

(a)

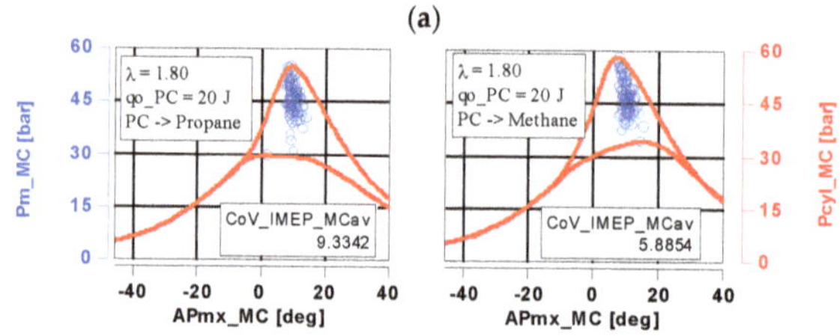

(b)

Figure 6. Stability of engine operation determined by maximum cylinder pressure Pmx (blue dots): (**a**) smallest CoV(IMEP) < 1% when burning methane and propane: λ = 1.5 Eqo_PC = 30 J; (**b**) largest CoV(IMEP) > 10% when burning methane and propane: λ = 1.8 Eqo_PC = 20 J.

A comparison of IMEP values at each point of engine operation is briefly shown in Figure 7. All curves are drawn in the same color, but those showing significant deviations are marked in blue or black. It can be seen from the data presented that the highest instability of operation occurs at λ = 1.8. Two curves significantly deviating from the stability criterion of CoV(IMEP) < 3.5% were recorded. The maximum values are 9.33% and 11.45%. Comparing the combustion of methane and propane, it was found that the combustion of methane significantly degrades the combustion process more (no combustion at qo = 10 J and λ = 1.8). At other operating points, the combustion conditions are similar (at the same fuel doses qo_PC), i.e., combustion is deteriorated.

The engine's so-called "work maps" stability is also included in the $IMEP_n$-$IMEP_{n+1}$ coordinates (Figure 8). They indicate the variability of sequential engine cycles and illustrate in detail the changes in the cyclicality of engine operation. At the smallest value of the excess air ratio λ = 1.3, the greatest irregularity occurs at high fuel doses to the PC during the combustion of methane in the prechamber (Figure 8b—left graph). The variation averages ΔIMEP = 0.8 bar/cycle. The most stable engine operation occurs at λ = 1.5. The cyclic changes from cycle to cycle are less than ΔIMEP = 0.5 bar/cycle. The highest irregularity was observed at λ = 1.8 during propane combustion (Figure 8a). The ΔIMEP changes are almost 3 bar/cycle. During methane combustion, the ΔIMEP value is a maximum of 1.5 bar/cycle.

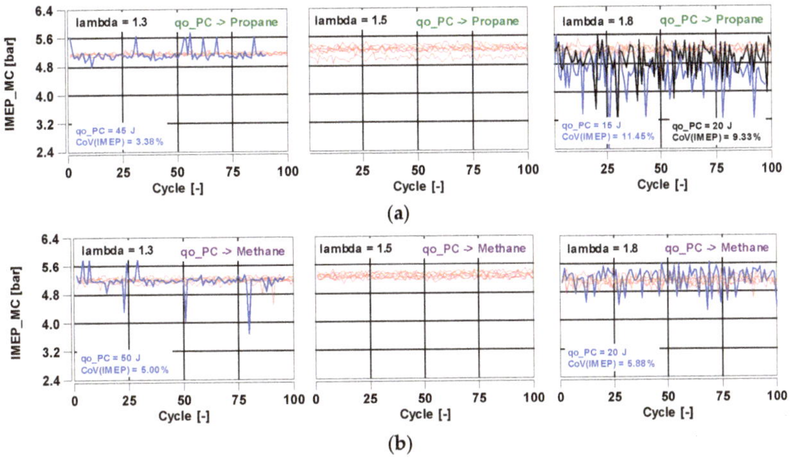

Figure 7. Instability of engine operation determined by IMEP for each value of excess air ratio: (**a**) when burning propane in the prechamber; (**b**) when burning methane in both chambers (values of the largest IMEP changes are marked in the figure).

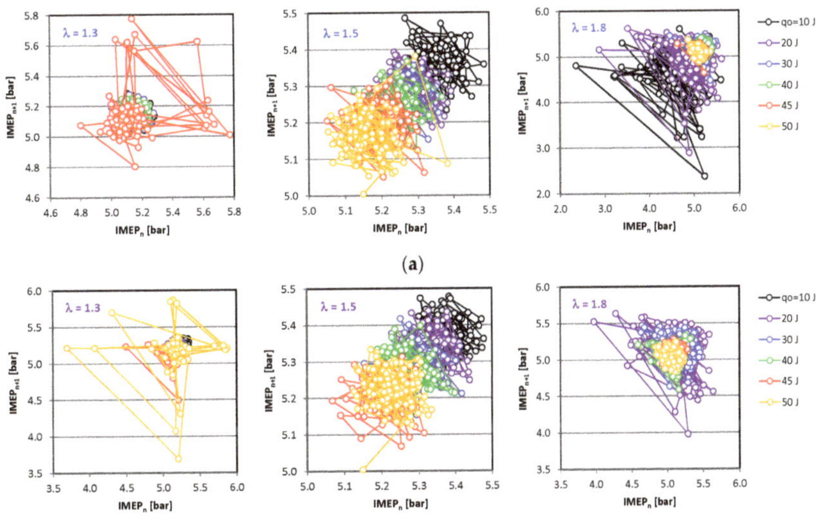

Figure 8. Unevenness of engine cycles determined by IMEP maps for each value of excess air ratio and different values of fuel dose energy injected into the prechamber: (**a**) when burning propane in the prechamber; (**b**) when burning methane in both chambers.

The analysis of inter-chamber tides is shown in Figure 9. The pressure difference between chambers was chosen as the parameter representing mass transfer intensity. The combustion of fuels at $\lambda = 1.3$ (Figure 9a) shows the smallest pressure difference between volumes when various fuels are combusted. A higher value was recorded during the combustion of propane in the prechamber, the pressure difference between the main and prechamber being about 0.9 bar. An earlier ignition of methane than propane is observed to achieve CoC = 8 deg aTDC. This means that propane combustion occurs faster, especially in the range of the first phase of combustion. This phase lasts from the beginning of combustion until 50% of the heat is released. When burning fuels at $\lambda = 1.5$, the start of combustion in the PC is slightly later for both fuels (Figure 9b). The pressure differences in the prechamber are already greater at about 1.4 bar. The combustion pressure in the PC during propane fueling is higher, and again the maximum falls slightly later than during methane combustion. The largest differences were observed when sparging the fuels at λ = 1.8 (Figure 9c). Differences in the onset of combustion in the PC are large at about 4 deg. Pressure differences in PC are about 2.1 bar.

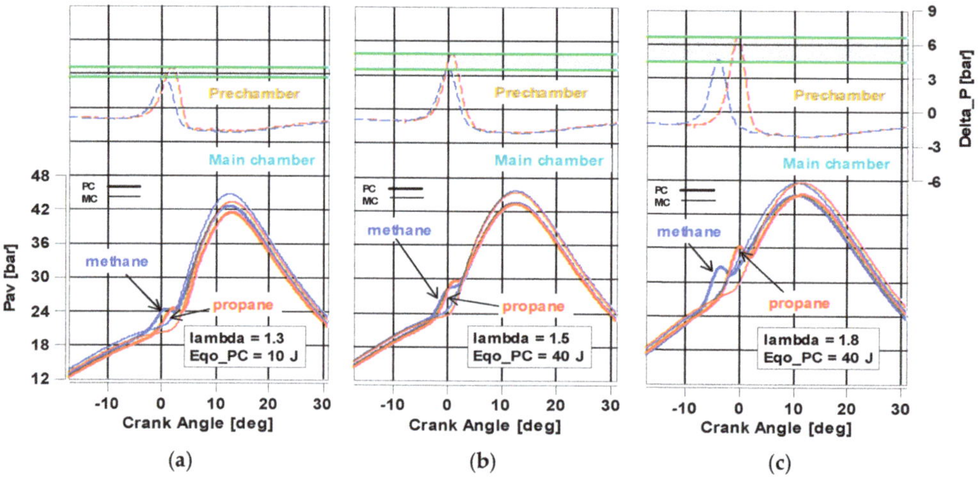

Figure 9. Inter-chamber flows (Delta_P) and pressure in the prechamber and main chamber during combustion of methane (blue line) and propane (red line): (**a**) during combustion of fuels with excess air λ = 1.3; (**b**) during combustion of fuels with excess air λ = 1.5; (**c**) during combustion of fuels with excess air λ = 1.8.

At all points, the maximum pressure in the MC is several bars higher than in the PC. This is mainly due to the throttling effect of the flow orifices produced in the prechamber.

The specific changes in the pressure values in the two chambers are shown in Figure 10. It shows the high intensity of the processes in the prechamber under the conditions of the mean energy supplied to this chamber. On this basis, it is possible to conclude the optimal amount of fuel delivered to the prechamber. As can be seen from the data presented, the best value is the energy in the range of 20–30 J delivered to the PC. Too small as well as too large a dose of fuel to the PC results in a non-intensive combustion in the prechamber. Regardless of the dose, higher pressure values were always recorded in the MC than in the PC.

Based on the above considerations of engine stability, the average values of IMEP in both combustion chambers were determined (Figure 11). It was found that the IMEP in MC has about 0.2 bar higher values than in PC. As the dose to the PC increases, the value decreases almost linearly. The change in IMEP is about 0.2 bar per 50 J of energy delivered to the PC. This decrease may be due mainly to the lower energy of the fuel contained in the main chamber. This means that the minimum dose delivered to the PC is sufficient to initiate the combustion process and maximize IMEP. The figures also show the effect of the loss of stability of engine operation, which is a very large drop in IMEP at λ = 1.8 and low energy fed to the PC. With such conditions of high excess air, the minimum dose given to the PC is too low to achieve proper combustion.

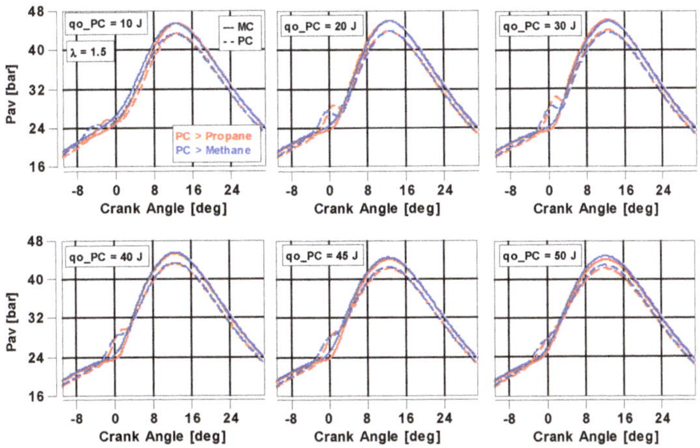

Figure 10. Conditions of pressure change in the cylinder during the combustion of propane (PC) and methane (MC), as well as combinations of methane (PC) and methane (MC) at different values of energy injected into the prechamber at λ = 1.5 (the most favorable combustion conditions regardless of the fuel initial dose).

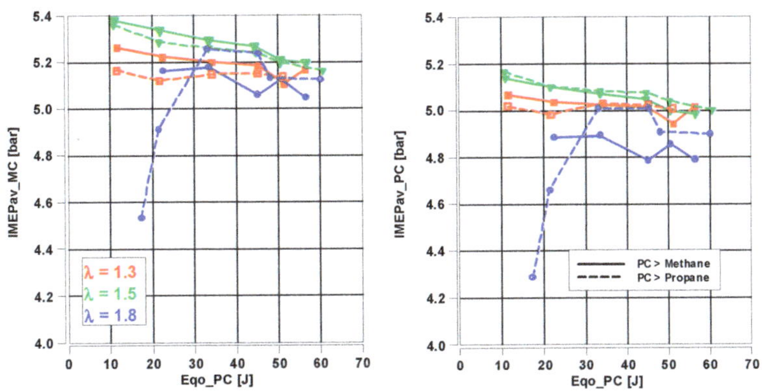

(a) (b)

Figure 11. Variation in IMEP with respect to energy delivered to the prechamber at different values of excess air ratio λ: (**a**) variation in IMEP in the main chamber; (**b**) variation in IMEP in the prechamber.

The combustion of methane and propane at λ = 1.8 is not beneficial. In both cases, an unsatisfactory combustion was obtained at small values of qo_PC. At larger values of qo_PC, the smallest IMEP values were observed, indicating that the excess air ratio limit was exceeded with the combustion system used.

Based on the pressure curve in the cylinder and Equation (3), the rate of heat release was determined. Integrating these values, the total amount of heat released was obtained. The analysis of this quantity in Figure 12 confirms the above information about the minimization of the dose delivered to the PC.

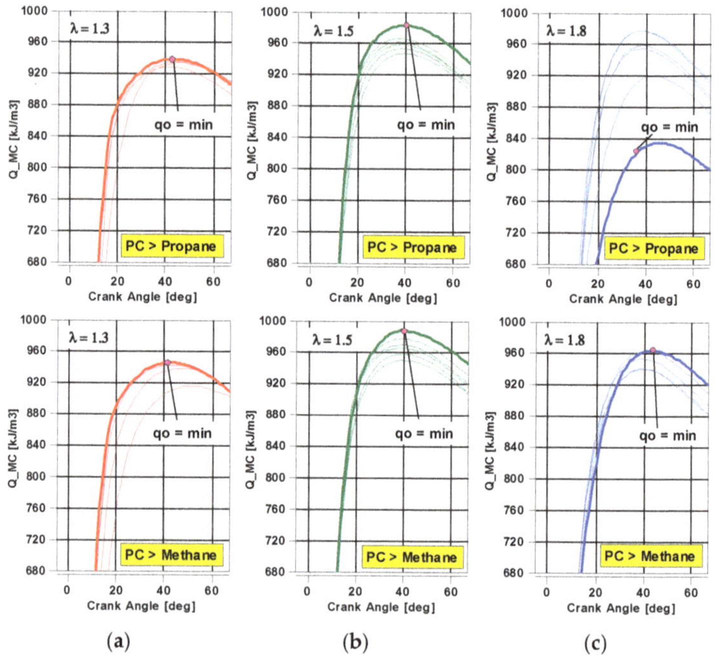

Figure 12. The heat release path in the cylinder (in the main chamber) with the determination of the minimum dose of fuel injected into the prechamber for both types of fuels: (**a**) at λ = 1.3; (**b**) at λ = 1.5; (**c**) at λ = 1.8.

The combustion of propane and methane in the PC results in the maximum heat release values being obtained with the minimum fuel dose. The combustion of very lean mixtures (λ = 1.8—Figure 12c) results in a different process when methane or propane is injected into the PC. When burning propane, the minimum dose is too low to achieve proper combustion, and the amount of heat released is the smallest. In this case, very large spreads in the path of heat release were obtained, depending on the amount of energy delivered to the PC. The combustion of methane (Figure 12c) also results in a maximum of heat released at qo → min, i.e., 10 J of energy delivered to the PC. However, the amount of heat released is at the same time the smallest (compared to other values λ).

4.3. Indicators of Engine Operation

4.3.1. Thermodynamic Indicators

Using the quantities in Figure 12 and taking into account Equation (1), the thermodynamic indices of an engine fueled by methane and propane were determined. The combustion center (CoC) and start (AI05) and end of combustion (AI90) were determined. The results of these measures are included in Figure 13.

According to data from Figure 13, the combustion of methane and propane in PC at λ = 1.3 and 1.5 results in high operation stability. As the energy delivered to the PC increases, the time of combustion I phase also increases. This is due to the control method of keeping the CoC constant (8 deg aTDC). Such control requires an increase in the ignition advance value, which results in an advance at the start of combustion. At the same time, during the combustion of both fuels, a longer combustion phase II (AI90—CoC) is also observed. This means that the fastest process occurs when small doses of fuel are fed into the prechamber. The combustion of fuels at λ = 1.8 results in large variations in thermodynamic indicators. This is due to the combustion conditions (operating instability) presented previously. Although the CoC value was kept constant, both the start and the end of combustion have a different trend from the previous values of the excess air ratio.

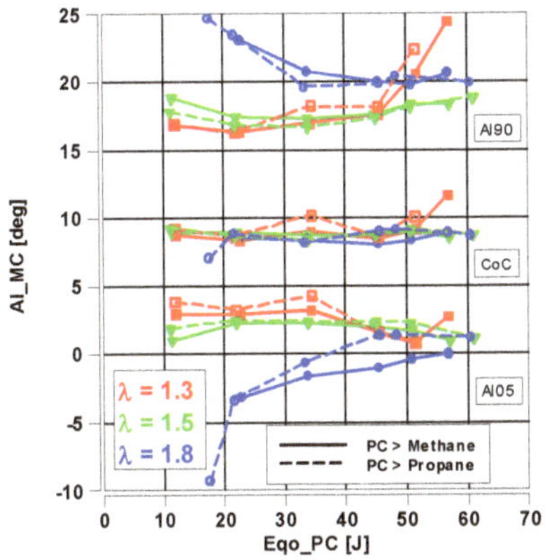

Figure 13. Thermodynamics of the combustion process: beginning of combustion (AI05), center of combustion (CoC), and end of combustion (AI90) determined for average pressure curves during methane and propane combustion.

4.3.2. System Efficiency for Methane and Propane Combustion

Taking into account Equations (4)–(6), the efficiency of the engine was determined at each operating point when the prechamber was fed with methane and propane (Figure 14). It was found earlier that the operating conditions at $\lambda = 1.8$ were not fully acceptable. Due to stable operating conditions, the highest engine efficiency was observed at $\lambda = 1.5$ when the prechamber was fueled with methane. The values of ηi reach more than 40% at a low fuel initial dose. Slightly lower values were observed at this operating point when PC was fueled with propane ($\eta_i = 39.3\%$). The combustion of mixtures with $\lambda = 1.3$ results in an efficiency slightly lower than at $\lambda = 1.5$, but higher than at $\lambda = 1.8$, at low doses of fuel delivered to the prechamber.

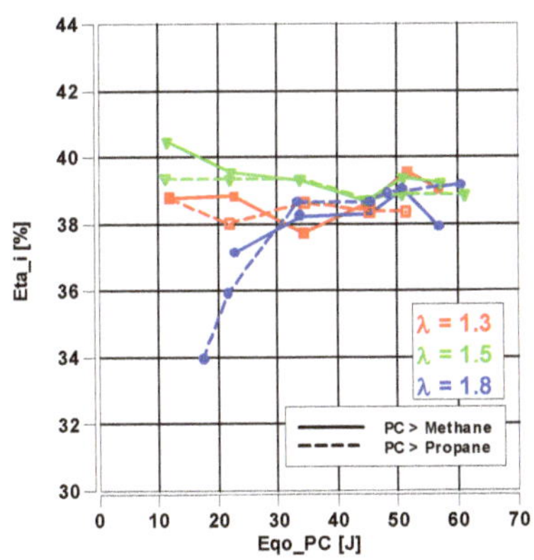

Figure 14. Indicated engine efficiency related to the value of energy supplied to the prechamber during the combustion of methane and propane.

Prechamber methane combustion is more beneficial at small fuel doses to PC and at $\lambda = 1.3$ and 1.5 in the range up to 30 J of energy in PC. The areas of increased engine efficiency for propane combustion in a two-stage system cannot be clearly identified.

Based on the above relationships, interpolated maps of the indicated engine efficiency fueled by propane and methane to PC were determined (Figure 15). As can be seen from the data presented, there are areas of higher efficiency when burning methane in PC than when burning propane. The combustion of methane in the PC results in higher efficiency for both small and large doses of this fuel fed to the PC. The combustion of small doses of propane in the PC at $\lambda = 1.8$ results in combustion efficiency being rapidly reduced. Similar negative engine operating conditions were noted when burning methane (no measuring point).

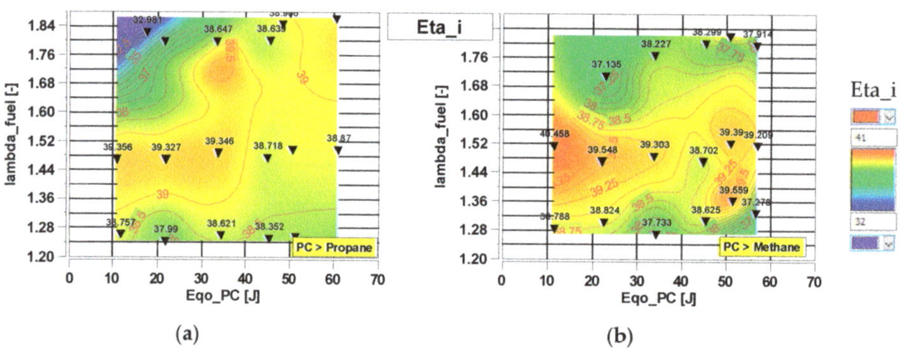

Figure 15. Indicated engine efficiency maps related to the energy contained in the fuel dose to the prechamber: (**a**) propane; (**b**) methane.

4.3.3. Analysis of Emission Indexes

During engine operation, the concentrations of carbon monoxide CO, hydrocarbons HC, and nitrogen oxide NO were analyzed, which were then converted into specific emissions relative to the power generated by the engine, and the results are shown in Figures 16–18. The main factor affecting emissions is the lambda excess air factor, and the trends obtained are consistent with the results presented in another paper [29].

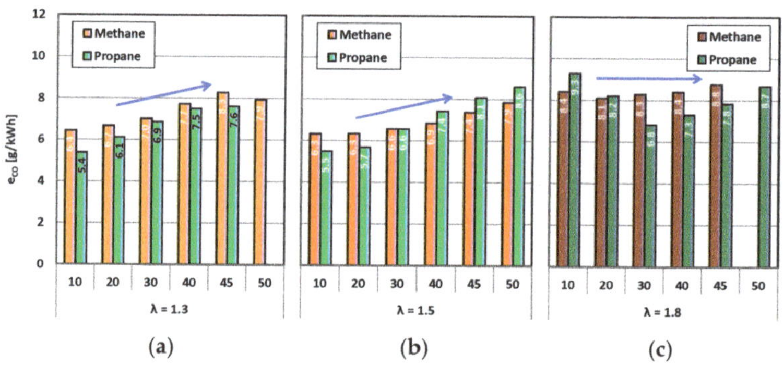

Figure 16. Carbon monoxide emissions at different values of energy supplied to the prechamber: (**a**) $\lambda = 1.3$; (**b**) $\lambda = 1.5$; and (**c**) $\lambda = 1.8$ (the arrows indicate the trend).

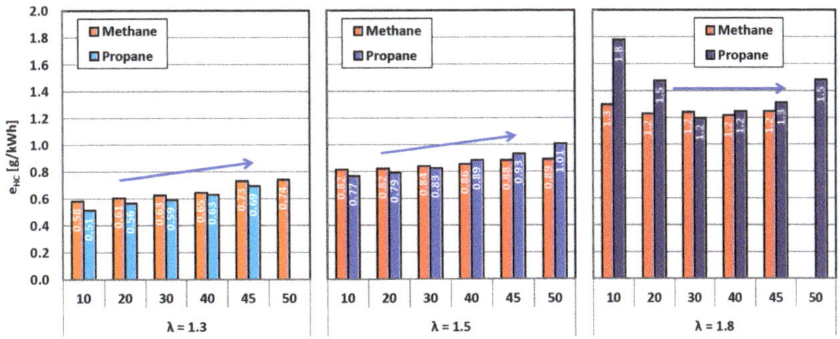

(a) (b) (c)

Figure 17. Hydrocarbon emissions at different values of energy supplied to the prechamber: (**a**) λ = 1.3; (**b**) λ = 1.5; and (**c**) λ = 1.8 (the arrows indicate the trend).

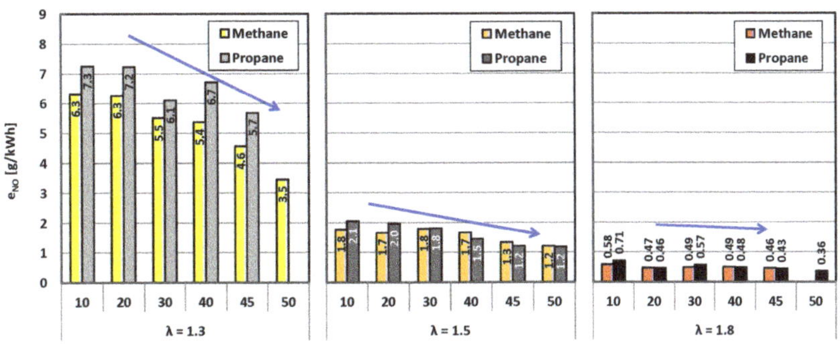

(a) (b) (c)

Figure 18. Nitrogen oxide emissions at different values of energy supplied to the prechamber: (**a**) λ = 1.3; (**b**) λ = 1.5; and (**c**) λ = 1.8 (the arrows indicate the trend).

The specific carbon monoxide CO emissions generated due to incomplete combustion, among other factors, are shown in Figure 16. For charges with an excess air ratio of 1.3, regardless of the amount of fuel delivered to the ignition chamber, lower emissions were generated by injecting propane into the PC. The largest differences of 1 g/kWh were achieved for the minimum energy in the PC. This may be due to more intensive ignition processes of the main charge determined by the flow of charge between chambers when combustion starts in the PC. Subsequently, increasing the charge dilution, i.e., λ = 1.5, reverses the trend of lower CO for propane as the fuel initiating combustion except for the two smallest values of the share of energy delivered to the PC. For ultra-lean charges of λ = 1.8, emissions for the single-fuel mode were more or less constant regardless of the amount of fuel delivered to the PC, which correlates with the stability of engine operation in this area. Using propane for lower fuel doses to the PC where the engine operated unstably resulted in higher emissions, which decreased sequentially as the proportion of the dose to the PC increased.

The next step analyses hydrocarbon emissions from unburned fuel and lubricating oil (Figure 17). Increasing the proportion of air in the mixture promotes increased HC emissions due to the deterioration of the combustion process and the incompletion of the flame in all areas of the combustion chamber. In the case of excess air ratio λ = 1.3 and λ = 1.5, emissions increase as the proportion of fuel to PC increases, i.e., the main charge becomes leaner. In this case, the primary factor determining combustion efficiency throughout the cylinder volume is the lower ignition energy requirements of the main charge rather than the amount of energy supplied to the initial combustion chamber. Increasing the excess air ratio in the leanest area decreases the importance of the fuel dose size to PC, especially for single-fuel operation. Propane supply to PC is better for lean charges, while for ultra-lean charges, single-fuel supply is better.

Nitrogen oxide NO emissions depend mainly on peak temperatures in the cylinder during the combustion process [30]. Dilution of the charge causes a decrease in the mentioned temperature, which leads to a decrease in NO emissions with an increase in the lambda excess air ratio. In the cases analyzed, NO emissions decrease with an increase in the proportion of fuel supplied to the PC. This is due to the dilution of the main charge mostly responsible for the emission of toxic exhaust components. In most cases, better NO emission rates were obtained for the single-fuel mode (Figure 18), corresponding to higher thermal efficiency, i.e., improving the combustion process due to energy indicators.

5. Conclusions

As a result of the experimental work carried out with a single-cylinder engine operating in single-fuel (methane) and dual-fuel methane (MC) + propane (PC) modes, it was found that

1. Feeding the engine in a single-fuel mode methane (MC) + methane (PC) allows better thermodynamic parameters of the combustion process to be obtained in relation to the dual-fuel mode where propane is supplied to the ignition chamber instead of methane.
2. The effect of dividing the fuel dose between the MC and PC chambers has the greatest impact in the case of combustion of mixtures with excess air ratios of 1.3 and 1.5.

The advantages of a single-fuel methane (MC) + methane (PC) system over a dual-fuel methane (MC) + propane (PC) system are as follows:

1. Better stability of engine operation CoV(IMEP) < 3.5% in the range of small doses of fuel fed to the prechamber and at charges with $\lambda = 1.3–1.5$.
2. Higher values of indicative mean effective pressure IMEP when burning mixtures with excess air ratio in the range $\lambda = 1.3–1.5$; these values are slightly higher when measured in the main chamber than in the prechamber.
3. Higher maximum values of heat released; at the same time, the combustion of minimum doses in the prechamber resulted in maximum values of heat released in relation to the other values of energy supplied to the PC.
4. Better thermodynamic performance of combustion in the prechamber at small doses fed to the PC and at $\lambda = 1.3$ and 1.5 in the range up to 30 J. The areas of increased propane combustion efficiency in the two-stage system cannot be clearly identified.
5. Higher value of the engine's indicated efficiency reaching more than 40% at a small dose of fuel to the prechamber against 39.3% when propane is fed to the ignition chamber.

The positive aspects of burning propane in a prechamber are as follows:

1. Shorter combustion time; the combustion time increases with an increase in λ and with an increase in the dose delivered to the prechamber.
2. Ignitability of the mixture at high λ and low fuel doses (no ignitability when using methane).
3. Reduction in CO and HC emissions when burning leaner loads.

Author Contributions: Conceptualization, I.P. and F.S.; methodology, I.P. and F.S.; software, I.P. and F.S.; validation, I.P. and F.S.; formal analysis, I.P. and F.S.; investigation, I.P. and F.S.; resources, I.P. and F.S.; data curation, I.P. and F.S.; writing—original draft preparation, I.P. and F.S.; writing—review and editing, I.P. and F.S.; visualization, I.P. and F.S.; supervision, I.P. and F.S.; project administration, I.P. and F.S.; funding acquisition, I.P. and F.S. All authors have read and agreed to the published version of the manuscript.

Funding: This research received no external funding.

Data Availability Statement: Not applicable.

Conflicts of Interest: The authors declare no conflict of interest.

References

1. Gracz, W.; Marcinkowski, D.; Golimowski, W.; Szwajca, F.; Strzelczyk, M.; Wasilewski, J.; Krzaczek, P. Multifaceted comparison efficiency and emission characteristics of multi-fuel power generator fueled by different fuels and biofuels. *Energies* **2021**, *14*, 3388. [CrossRef]
2. Davis, G.C.; Krieger, R.B.; Tabaczynski, R.J. *Analysis of the Flow and Combustion Processes of a Three-Valve Stratified Charge Engine with a Small Prechamber*; SAE Technical Paper; SAE International: Warrendale, PA, USA, 1974. [CrossRef]
3. Adams, T. *Torch Ignition for Combustion Control of Lean Mixtures*; SAE Technical Paper; SAE International: Warrendale, PA, USA, 1979. [CrossRef]
4. Alvarez, C.E.C.; Couto, G.E.; Roso, V.R.; Thiriet, A.B.; Valle, R.M. A review of prechamber ignition systems as lean combustion technology for SI engines. *Appl. Therm. Eng.* **2018**, *128*, 107–120. [CrossRef]
5. Pielecha, I.; Szwajca, F. Lean methane mixtures in turbulent jet ignition combustion system. *Energies* **2023**, *16*, 1236. [CrossRef]
6. Zhu, S.; Akehurst, S.; Lewis, A.; Yuan, H. A review of the pre-chamber ignition system applied on future low-carbon spark ignition engines. *Renew. Sustain. Energy Rev.* **2022**, *154*, 111872. [CrossRef]
7. Allison, P.M.; de Oliveira, M.; Giusti, A.; Mastorakos, E. Pre-chamber ignition mechanism: Experiments and simulations on turbulent jet flame structure. *Fuel* **2018**, *230*, 274–281. [CrossRef]
8. Wu, H.; Wang, L.; Wang, X.; Sun, B.; Zhao, Z.; Lee, C.-F.; Liu, F. The effect of turbulent jet induced by pre-chamber sparkplug on combustion characteristics of hydrogen-air pre-mixture. *Int. J. Hydrogen Energy* **2018**, *43*, 8116–8126. [CrossRef]
9. Gholamisheeri, M.; Wichman, I.S.; Toulson, E. A study of the turbulent jet flow field in a methane fueled turbulent jet ignition (TJI) system. *Combust. Flame* **2017**, *183*, 194–206. [CrossRef]
10. Liu, Z.; Zhou, L.; Zhong, L.; Wei, H. Enhanced combustion of ammonia engine based on novel air-assisted pre-chamber turbulent jet ignition. *Energy Convers. Manag.* **2023**, *276*, 116526. [CrossRef]
11. Onofrio, G.; Napolitano, P.; Tunestål, P.; Beatrice, C. Combustion sensitivity to the nozzle hole size in an active pre-chamber ultra-lean heavy-duty natural gas engine. *Energy* **2021**, *235*, 121298. [CrossRef]
12. Zhou, L.; Liu, P.; Zhong, L.; Feng, Z.; Wei, H. Experimental observation of lean flammability limits using turbulent jet ignition with auxiliary hydrogen and methane in pre-chamber. *Fuel* **2021**, *305*, 121570. [CrossRef]
13. Bueschke, W.; Szwajca, F.; Wislocki, K. *Experimental Study on Ignitability of Lean CNG/Air Mixture in the Multi-Stage Cascade Engine Combustion System*; SAE Technical Paper; SAE International: Warrendale, PA, USA, 2020. [CrossRef]
14. Vera-Tudela, W.; Merotto, L.; Balmelli, M.; Soltic, P. Experimental study of the ignition of lean methane/air mixtures using inductive and NRPD ignition systems in the pre-chamber and turbulent jet ignition in the main chamber. *Energy Convers. Manag.* **2022**, *252*, 115012. [CrossRef]
15. Arslan, E.; Kahraman, N. Comparison of natural gas and propane addition to combustion air in terms of engine performance in compression ignition engine. *Fuel* **2022**, *312*, 122952. [CrossRef]
16. Jeong, S.Y.; Jang, D.; Lee, M.C. Property-based quantitative risk assessment of hydrogen, ammonia, methane, and propane considering explosion, combustion, toxicity, and environmental impacts. *J. Energy Storage* **2022**, *54*, 105344. [CrossRef]
17. Li, X.; He, B.-Q.; Zhao, H. Effect of direct injection dimethyl ether on the micro-flame ignited (MFI) hybrid combustion characteristics of an optical gasoline engine at ultra-lean conditions. *Fuel Process. Technol.* **2020**, *203*, 106383. [CrossRef]
18. Distaso, E.; Amirante, R.; Cassone, E.; De Palma, P.; Sementa, P.; Tamburrano, P.; Vaglieco, B.M. Analysis of the combustion process in a lean-burning turbulent jet ignition engine fueled with methane. *Energy Convers. Manag.* **2020**, *223*, 113257. [CrossRef]
19. Liu, Z.; Zhou, L.; Wei, H. Experimental investigation on the performance of pure ammonia engine based on reactivity controlled turbulent jet ignition. *Fuel* **2023**, *335*, 127116. [CrossRef]
20. Vinod, K.N.; Fang, T. Experimental characterization of spark ignited ammonia combustion under elevated oxygen concentrations. *Proc. Combust. Inst.* **2022**. [CrossRef]
21. Zhang, X.; Tian, J.; Cui, Z.; Xiong, S.; Yin, S.; Wang, Q.; Long, W. Visualization study on the effects of pre-chamber jet ignition and methane addition on the combustion characteristics of ammonia/air mixtures. *Fuel* **2023**, *338*, 127204. [CrossRef]
22. Liu, Z.; Zhou, L.; Zhong, L.; Wei, H. Reactivity controlled turbulent jet ignition (RCTJI) for ammonia engine. *Int. J. Hydrogen Energy* **2023**, *48*, 12519–12522. [CrossRef]
23. Lee, M.C.; Seo, S.B.; Yoon, J.; Kim, M.; Yoon, Y. Experimental study on the effect of N_2, CO_2, and steam dilution on the combustion performance of H_2 and CO synthetic gas in an industrial gas turbine. *Fuel* **2012**, *102*, 431–438. [CrossRef]
24. Cheng, Q.; Kaario, O.; Ahmad, Z.; Vuorinen, V.; Larmi, M. Effect of pilot fuel properties on engine performance and combustion stability in a tri-fuel engine powered by premixed methane-hydrogen and diesel pilot. *Int. J. Hydrogen Energy* **2021**, *46*, 37469–37486. [CrossRef]
25. Jean, M.; Granier, P.; Leroy, T. Combustion stability control based on cylinder pressure for high efficiency gasoline engines. *Energies* **2022**, *15*, 2530. [CrossRef]
26. Heywood, J.B. *Internal Combustion Engine Fundamentals*; McGraw-Hill: New York, NY, USA, 1988.

27. Batool, S.; Naber, J.D.; Shahbakhti, M. Data-driven modeling and control of cyclic variability of an engine operating in low temperature combustion modes. *IFAC-PapersOnLine* **2021**, *54*, 834–839. [CrossRef]
28. Merkisz, J.; Pielecha, J.; Radzimirski, S. *New Trends in Emission Control in the European Union*; Springer Tracts on Transportation and Traffic; Springer: Cham, Switzerland, 2014; Volume 4. [CrossRef]
29. Lott, P.; Deutschmann, O. Lean-burn natural gas engines: Challenges and concepts for an efficient exhaust gas aftertreatment system. *Emiss. Control Sci. Technol.* **2020**, *7*, 1–6. [CrossRef]
30. Li, G.; Long, Y.; Zhang, Z.; Liang, J.; Zhang, X.; Zhang, X.; Wang, Z. Performance and emissions characteristics of a lean-burn marine natural gas engine with the addition of hydrogen-rich reformate. *Int. J. Hydrogen Energy* **2019**, *44*, 31544–31556. [CrossRef]

energies
MDPI

Article

Effect of a Cu-Ferrite Catalyzed DPF on the Ultrafine Particle Emissions from a Light-Duty Diesel Engine

Eugenio Meloni [1,*], Bruno Rossomando [1], Gianluigi De Falco [2], Mariano Sirignano [2], Ivan Arsie [3] and Vincenzo Palma [1]

[1] Department of Industrial Engineering, University of Salerno, Via Giovanni Paolo II 132, 84084 Fisciano, Italy
[2] Department of Chemical, Materials and Production Engineering, University of Naples "Federico II", P.le Tecchio 80, 80125 Napoli, Italy
[3] Department of Engineering, University of Naples "Parthenope", Centro Direzionale Isola C4, 80143 Napoli, Italy
* Correspondence: emeloni@unisa.it

Abstract: The emissions of diesel engines in terms of particulate matter are limited all over the world. One possible solution for reaching the target imposed by the various regulations could be the adoption of a catalytic diesel particulate filter (CDPF). Nevertheless, the effect of CDPFs on the particle size distributions (PSDs) during the regeneration process needs to be deeply investigated. Therefore, this research work is focused on a detailed PSD analysis during the active regeneration of a 30 %wt $CuFe_2O_4$ loaded CDPF at the exhaust of an L-D diesel engine to reach a more complete understanding of the filter behavior. The results of the experimental tests evidence that at the CDPF outlet, compared to a standard DPF: (i) during the start-up of the regeneration, the particle emissions are three orders of magnitude lower and remain two orders of magnitude lower for particle sizes larger than 50 nm; (ii) the PSDs measured in the time range of 200–450 s exhibit the bimodality observed during the accumulation phase, with a peak that is three orders of magnitude lower; (iii) at the end of the regeneration, the PN distribution exhibits reductions of two and three orders of magnitude for particle sizes of 5 nm and above 50 nm, respectively.

Keywords: diesel engine soot abatement; particulate matter; catalytic diesel particulate filter; particle size distribution; filter regeneration

Citation: Meloni, E.; Rossomando, B.; De Falco, G.; Sirignano, M.; Arsie, I.; Palma, V. Effect of a Cu-Ferrite Catalyzed DPF on the Ultrafine Particle Emissions from a Light-Duty Diesel Engine. *Energies* **2023**, *16*, 4071. https://doi.org/10.3390/en16104071

Academic Editors: Monika Kosowska-Golachowska and Tomasz Czakiert

Received: 5 April 2023
Revised: 26 April 2023
Accepted: 11 May 2023
Published: 13 May 2023

1. Introduction

Diesel engines are still considered high-efficiency energy conversion systems since their operation under fuel lean combustion allows the production of lower CO_2 [1], carbon monoxide (CO) and hydrocarbon (HC) emissions compared to spark ignition engines [2]. On the other hand, other compounds are present in the exhaust emissions of diesel engines, both in the gas phase and in the solid phase as particles [1]. The gas phase is commonly characterized by the presence of N_2, CO_2, O_2, H_2O, CO and NOx, as the main components, and the well-known soot is the main component of the solid phase [3]. In the last few years, the use of biodiesel and metallic additives dispersed in diesel have been proved to improve engine efficiency and reduce emissions [4]. However, the former introduces new challenges, such as corrosion in metal alloys, aluminum, and copper [4]. In any case, diesel engines are characterized by high particulate matter (PM) emissions, since some of their peculiarities, in particular, their fuel composition and the operating conditions realized in the combustion chamber, allow nonhomogeneous mixing, resulting in favored PM formation in the high-temperature regions with higher fuel contents [2]. PM is constituted of volatile and aromatic compounds (polycyclic aromatic hydrocarbons (PAHs) and heavier aromatic compounds) and non-volatile compounds (soot) [5]. Soot is characterized by a turbostratic structure, in which the particles are marked by the presence of crystallites of discernable lengths and with some stacking faults in their peripheries, while their centers have random crystallite orientations and are highly disordered [6]. Moreover, a thin-layer SOF (soluble organic fraction) may be formed on these particles, since small amounts of fuel and engine oil from the crankcase may escape oxidation reactions in the combustion chamber. It is well known that diesel emissions have direct negative effects on human health, causing several kinds of damage, including asthma, lung and heart disease, and cancer, and the main carcinogenic and mutagenic respiratory health risks are related to exposure to PAH, in gas phase and/or adsorbed on soot [7,8]. These harmful effects have led to the limitation of PM emissions from diesel engines all over the world through proper regulations, with the limits becoming more stringent over time [9]. For example, particle mass limits have dropped by more than 95% from the early 1990s (Euro I) to 2009 (Euro V) for light-duty diesel (L-D) vehicles in Europe [7]. A particle number (PN) limit (6×10^{11} particles/km, taking into consideration only particles over 23 nm) was first introduced in 2011 (Euro Vb) for L-D diesel vehicles and was introduced in EURO VI emission standards also for gasoline direct injection (GDI) engines [10,11]. Car manufacturers have widely introduced diesel particulate filters (DPFs) in their vehicles to comply with these limits; thus, DPFs have become a crucial technology for achieving the tradeoff between emission limits and engine performance. In general, the DPF is a ceramic monolith with a honeycomb multi-channel structure composed of silicon carbide (SiC), whose parallel channels are alternately plugged, creating the so-called wall-flow configuration. In this way, the exhaust gas is constrained to go through the porous walls of the channels [12]. These devices ensure PM reduction with a quite impressive efficiency—more than 95% in terms of mass and concentrations higher than 99% over a wide range of engine operating conditions [12,13]. Since the particles are trapped by the millions of complex structured pores inside the DPF structure via the physical filtration principle, either the periodic or the continuous elimination of these particles is key to ensuring a long lifetime of the filter [14]. This removal process is called regeneration and consists in the oxidation of the trapped soot [15]. Efficient regeneration is required in order to keep the fuel penalty due to DPF presence as low as possible. In fact, a regeneration process that is not able to fully burn the deposited soot may result in a fast increase in the backpressure, consequently causing different drawbacks, including increased fuel consumption, engine failure or even fire events [16]. During the regeneration step, only the organic components can be burned, while the salts, which are known as DPF ash, survive and require manual dredging. For a diesel vehicle driven 20,000 km per year, regeneration would be expected every 4–14 days if a frequency of regeneration events is assumed to vary from 250 to 800 km [17]. In recent years, the development of models that are able to accurately predict DPF behavior in all its functioning phases (from the soot loading up to the regeneration stages) has become mandatory. Therefore, the attention of researchers has focused on this issue.

Three main regeneration strategies can be adopted, namely:

i. Active regeneration, in which the exhaust gas is heated up to the oxidation temperature of soot, about 550–600 °C [18–20];
ii. Passive regeneration, in which the soot oxidation temperature is lowered by the adoption of a catalyst [21,22]; and
iii. Composite regeneration, which consists of a combination of the other two [23].

Regarding the passive regeneration strategy, the catalyst can be either mixed with the fuel (fuel-borne catalyst, FBC) or deposited on the DPF, thereby obtaining a catalyzed DPF (CDPF) [7]. CeO_2 is the most widely used FBC; it can lower soot oxidation temperature and consequently soot emissions. However, recent studies have evidenced that its use reduces soot and THC emissions by up to 30% at part load under low-temperature combustion conditions, but no significant differences are observed at high load [24]. Moreover, the addition of CeO_2 and $Fe(C_5H_5)_2$ nanoparticles to ultralow sulfur diesel (ULSD) may increase the total particle count, due to the formation of self-nucleated metallic nanoparticles, and toxic effects on human health may occur due to modifications in the physicochemical characteristics of PM caused by FBC-doped fuels [25]. Apart from CeO_2, different metals have been proposed as fuel additives, including Mn, Fe, Cu, Al, Be and Pt, all showing positive effects in reducing diesel engine emissions [4]. In recent years, different active species have been proposed for use in soot oxidation, both in powder forms and deposited on DPFs, including $Ce_{0.5}Zr_{0.5}O_2$ catalysts promoted by multivalent transition-metal (Mn, Fe and Co) oxides [26]; nanostructured structures, such as equimolar ceria-praseodymia [27]; and CoOx-decorated CeO_2 ($CoCeO_2$) heterostructured catalysts [28], copper ferrite ($CuFe_2O_4$) [22], Ag/Al_2O_3 [29], nanoscale Mn_3O_4 [30], Mn_2O_3 [31], Fe-doped Mn_2O_3 [32] and CeO_2 [33]. All the proposed catalysts have shown good activity in soot burning, with soot temperature oxidation in the range of 300–500 °C.

DPF regeneration in L-D vehicles, both in active and passive strategies, is achieved by using an after/post-injection strategy to increase the temperature of the exhausts up to soot burnout temperature. In particular, this strategy involves an after- and a post-injection: the former allows, through the injection of a checked amount of fuel during the expansion stroke, the evaporation of the latter, which occurs during the exhaust valve opening [34]. The unburned and vaporized fuel exiting the combustion chamber is subsequently oxidized in the diesel oxidation catalyst (DOC). This secondary fuel oxidation allows the exhaust gas to reach the ignition soot temperature (500–600 °C) across the DPF. A deep understanding of the phenomena related to the post-injection allows extremely important improvements to DPF operation modes, also in terms of extra fuel penalties during active regeneration [35–37].

Although the DPF regeneration condition has not been included yet in vehicle emission regulations, many previous studies have emphasized the emissions that occur during regeneration, evidencing how gaseous pollutants, as well as particulate emissions, in terms of mass and number, largely increase [1,35]. Experimental tests performed with the aim of investigating active regeneration events evidenced remarkable transient particulate emissions. Various mechanisms have been proposed in the literature to explain these observations [9,36,38–42]:

i. A decreased filtration efficiency may be due to the particles being taken away from the cake layer and from inside the porous walls;
ii. The increased temperature due to active regeneration has been suggested to lead to expansion of the porous wall, resulting in a further lower filtration efficiency;
iii. The particle flow into the exhaust system may be changed due to different operating conditions in the engine in the case of active regeneration; fuel injection into the engine cylinder towards the end of the expansion stroke could change it significantly;
iv. The injection of extra fuel into the tailpipe may lead to the formation of additional particles upstream of the inlet of the DPF (the main hypothesis is that the trapped particles may fragment during rapid oxidation events, thereby escaping the DPF, resulting in particle breakthrough phenomena);
v. Active regeneration may be characterized by the emission of semi-volatile particles in addition to solid ones.

A deep understanding of the particle emissions during the active regeneration of a non-catalyzed wall-flow SiC DPF has been made possible by performing dedicated experimental tests [43]. In particular, a detailed analysis of PSDs at the exhaust of a EURO V L-D diesel engine was performed, with the engine working under different operating conditions, aiming at estimating the PN removal efficiency of the filter in the range of 5–100 nm. Moreover, the experimental tests allowed investigation of the evolution of PN concentration during the regeneration procedure, highlighting the influence of DPF temperature and the fuel post-injection strategy. The results of the PSD analysis have shown that:

i. At the DPF inlet, during the accumulation phase, a strong bimodality with sub-10 nm particles dominating the PN concentration is present;
ii. At the DPF outlet, a filtration efficiency higher than 90% was detected, demonstrating that the wall-flow filter can also remove the sub-23 nm particles. In particular, an efficiency value of about 99% has been detected in the range of 20–40 nm [43].

In previous works, we have demonstrated good results in terms of soot oxidation using a 30 %wt $CuFe_2O_4$ loaded catalytic DPF (CDPF) [23,44]. However, in the literature, the use of a CDPF resulted in high particle emissions, in particular during the first phase of the regeneration step [45]. Therefore, in this work the research was focused on detailed PSD analysis of the particles emitted during the active regeneration of the CDPF at the exhaust of a Euro V L-D diesel engine in order to reach a more complete understanding of the filter behavior, also aiming at its potential use in the real world. Using a scanning mobility particle sizer (SMPS), a PN distribution in the range of 4.5–160 nm was obtained during the soot filtration stage of the CDPF at various engine operating conditions. The filtration efficiency of the CDPF in the above specified range was calculated by alternately sampling the particles at the filter entrance and exit. Moreover, the regeneration process was deeply investigated in terms of the dynamic behavior of the PN size distributions.

2. Materials and Methods

2.1. The Monoliths

In this work, semicircular commercial monoliths in SiC (Pirelli Ecotechnology (Arese, Italy), 150 cpsi) with a radius of 9 cm were used to assemble the DPF. The bare monoliths were preliminarily calcined at 1000 °C for 48 h. This thermal treatment allowed the formation of a SiO_2 layer on SiC granules, increasing the adhesion of the active species over the filter even in the absence of a washcoat. Therefore, two filters with a total volume of about 3.5 L (D = 190 mm and L = 125 mm) were obtained by coupling four monoliths for each one, obtaining a V_{filter}/engine displacement ratio = 1.3. The main features of the assembled bare filters are shown in Table S1.

2.2. Catalytic Filter Preparation

The preparation procedure of the CDPF began with the preliminary dipping for 30 min of one of the assembled bare DPFs in a 1:1 HF:HNO_3 acid mixture maintained at 45 °C, aiming at increasing the initial average pore diameter [23]. This procedure allowed the enhancement of the catalyst loading capacity, ensuring similar pressure drops. The selected active species was copper ferrite ($CuFe_2O_4$), owing to its high activity and very good stability demonstrated in the catalytic oxidation of soot, shown in our previous works [10,22,23].

The catalyst precursor solution was prepared by mixing iron nitrate and copper nitrate (2:1 molar ratio) in distilled water, with continuous stirring at 80 °C. The CDPF was prepared following the previously optimized preparation procedure [44] by repeating the phases of impregnation in the prepared solution, with drying at 80 °C and calcination at 1000 °C after each impregnation, until the desired load of active species of 30 %wt was reached.

2.3. Characterization of the Catalytic Filters

Different characterization analyses were used for the characterization of the different materials employed in this work.

The active phase in powder form was investigated using the X-ray diffraction (XRD) technique (with a microdiffractometer (Rigaku (Tokyo, Japan) D-max-RAPID), using Cu-Kα radiation), aiming at verifying the formation of the desired crystalline forms of $CuFe_2O_4$.

The structured samples were characterized through:

i. Scanning electron microscopy (SEM), using a Philips Mod.XL30 coupled to an energy dispersive X-ray spectrometer (EDS; Oxford, UK) for the surface analysis;
ii. Hg porosimetry tests, with N2 adsorption at −196 °C (to calculate the samples' specific surface areas (SSAs) through BET);
iii. Ultrasound adherence test (exposure of the monoliths to ultrasound according to an optimized experimental procedure [44]).

Moreover, H_2-TPR measurements were performed to investigate the reducibility of the catalyst using a sample of the catalytic SiC monolith at room temperature up to 900 °C at a heating rate of 5 °C/min with a 5% H_2/N_2 flow. The concentrations of the gas stream were monitored by means of a mass spectrometer (HIDEN Analytical, Warrington, UK).

SEM images and Raman spectra (inVia Raman Microscope (Renishaw, Wotton-under-Edge, UK), equipped with a 514 nm Ar ion laser operating at 25 mW) were used for the characterization of the soot emitted by the diesel engine used for the experimental tests, as well as of the soot deposited on the catalytic filter after the deposition step.

2.4. Experimental Layout and Instruments

The experimental tests were carried out in the Energy and Propulsion Laboratory at the University of Salerno. The engine test bed, auxiliary plants, sensors and emission probes were in the engine test cell. The propulsion system was remotely controlled from an adjacent room, where the control console and the equipment for the management of the hardware, actuators and sensors were located. Figure 1 shows the engine test bed equipped with an eddy-current dyno (Borghi-Saveri, Bologna, Italy) and a light-duty EURO V common-rail diesel engine, whose main technical data are reported in Table S2.

Figure 1. Engine test bed.

The main hardware and software systems for the engine-dyno control are:

i. The AVL microIFEM for the management of the engine-dyno system by means of Puma open software;
ii. The AVL Indimicro to monitor and process the in-cylinder data by means of Indicom software;
iii. The control unit ETAS ES592.1 to manage the engine control unit (ECU) by means of INCA v7.1. The latter allows the visualization, processing and handling of the measurement signals and the control settings of the ECU, thus enabling engine operating modes different from the basic ECU calibration. In particular, it is possible to handle the combustion control variables, such as rail pressure, injection pattern, boost pressure and exhaust gas recirculation (EGR), to investigate specific engine operating conditions.

Focusing on the exhaust emissions, PM measurement was performed using the AVL Smoke Meter 415S (AVL List GmbH, Graz, Austria). The measurement of NO and NOx concentrations across the DPF was performed using the Cambustion CLD500 fast NOx sensor, a chemi-luminescence detector (CLD) with a response time of 3 ms.

The engine exhaust pipeline was modified in order to properly locate the DPFs and sensors. As schematically represented in Figure 2, the filter was placed downstream of the

DOC, and two variable section ducts were placed upstream and downstream of it. The CDPF was wrapped in a heat-expanding intumescent ceramic mat (Interam® Mat Mount 550 by 3M Italia Srl (Pioltello, Italy) before being enclosed in the stainless-steel reactor of the experimental plant. Temperature and pressure drop across the filter were continuously monitored by means of two K type thermocouples (at the inlet and outlet of the CDPF) and a differential pressure sensor (MDTP), respectively. The latter was composed of two sensors located upstream and downstream of the filter, and its signals were continuously processed by a microcontroller that returned the pressure drop. This device was powered in voltage by an external power supply set at 5 V. K thermocouples were connected to the AVL Puma Open, while the MTDP, after the calibration with a sphygmomanometer, was connected to a TLK. The output current signal was linked via RS232 to a Lab-View interface. For the investigation of passive regeneration with the CDPF, the measurement of NO and NO_2 concentrations was performed using the Cambustion CLD 500 (Cambustion, Cambridge, UK), equipped with two independent probes for NO and NOx sensing. An alternate sampling, arranged like the soot measurement described above, allowed investigation of the variation in NOx concentration across the filter [22]. The filter assembly in the exhaust pipe and the CDPF placed in the reactor are shown in Figure 3. The main sensors and measurement instruments are listed in Table S3, with the corresponding accuracies.

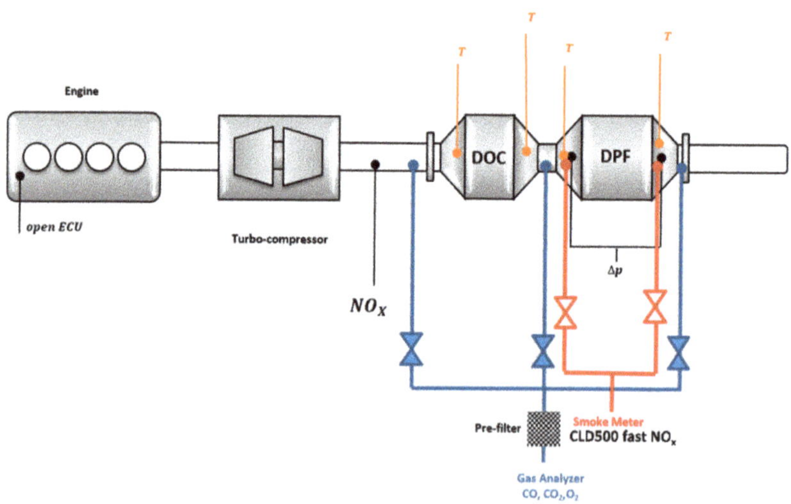

Figure 2. Schematic representation of the experimental layout and sensors.

The PSDs were measured by means of an SMPS system consisting of an electrostatic classifier (EC; TSI Model 3082 (TSI, Shoreview, MN, USA)), a nano-differential mobility analyzer (nano-DMA; TSI Model 3085 (TSI, Shoreview, MN, USA)) and an advanced aerosol neutralizer (TSI Model 3088 (TSI, Shoreview, MN, USA)) linked to a condensation particle counter (CPC; TSI Model 3750 (TSI, Shoreview, MN, USA)). A sample flow rate of 0.3 L/min and a sheath flow rate of 3 L/min in the EC were chosen as operating conditions, allowing the analysis of particle sizes in the range of 4.5–160 nm. For each sample, the data were reported as averaged values of multiple scans, where the duration of a single scan was set to 60 s. The PSD measurements were performed by diluting the aerosols sampled from the engine before entering the SMPS system. The dilution tool was a two-stage dilution system obtained from Dekati (fine particle sampler; Model FPS-4000 (Dekati, Kangasala, Finland)), and two dilution ratios (DRs) of 80 and 130 were used, depending on the operating condition. Moreover, the dilution temperature of the first stage was set at 25 °C. The dilution ratio was properly chosen in order to avoid the condensation/coagulation of particles along the sampling line. In other research papers, higher DRs were used for particle analysis, and heated lines were employed; however, it is worth noting that the specific experimental setup strongly influences the critical DR; the important parameters are PN, gas phase composition in the sampling line and residence time. The critical DR was checked in all the investigated conditions [46–48]. In this work, two different DRs were used to obtain, after the CDPF, a higher signal-to-noise ratio by using the lower DR, still achieving the critical dilution conditions.

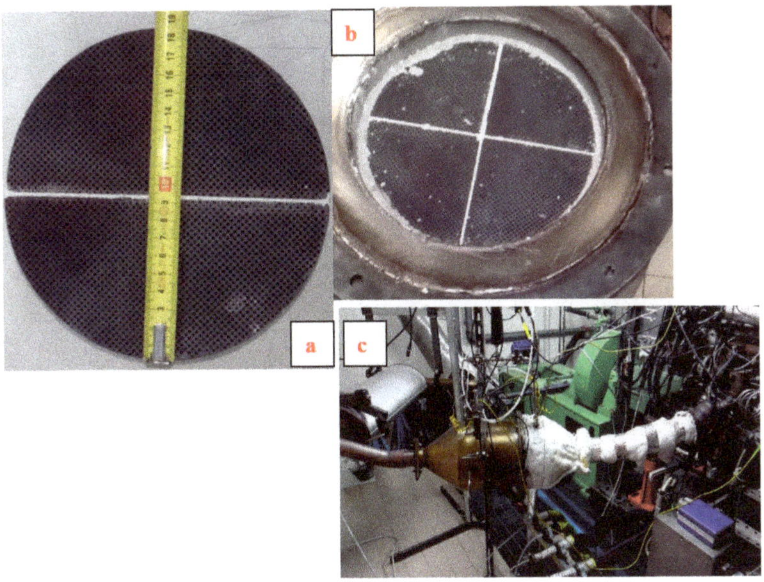

Figure 3. CDPF in the reactor used in the experimental campaign (**a**,**b**) and the filter assembly in the exhaust pipe (**c**).

3. Results and Discussion

3.1. SiC Monoliths before Catalyst Deposition

The dipping of the monoliths in the acid solution for 30 min before catalyst deposition had the effect of increasing both the initial average pore diameter from 17 to 22 μm and the SSA from 0.35 m^2/g to 2.20 m^2/g.

3.2. Soot Emitted by the Diesel Engine

The soot emitted by the diesel engine used for the experimental tests was characterized by means of SEM and SEM-EDX analysis (Figures S1 and S2).

The SEM images evidenced the typical spherical aggregates of soot, whose primary constituents had dimensions in the order of 50–300 nanometers. The SEM-EDX analysis evidenced that the elements detected on soot trapped by the CDPF during the experimental tests were the typical elements present in the fuel, namely, S, Cr, Mn and Ni. Other detected elements (Fe, Cu, Si and O) are the constituents of both carrier and active species.

The SEM analysis was used to also characterize the catalytic filter after the soot deposition stage in order to investigate the soot–catalyst contact (Figure 4). In particular, the following images of catalyst pieces after a soot deposition step were obtained, where a load of about 5 g/L was reached, and so at the same load of soot relevant to the regeneration step.

The SEM images shown in Figure 4 evidenced the soot–catalyst contact features. In particular, in Figure 4a it is possible to observe that the filter surface is entirely covered by the trapped soot with a very homogeneous distribution, confirming the very high CDPF ability to store a high soot amount with an optimal filtration efficiency. Moreover, Figure 4b,c are related to the inner surfaces of the carrier walls. It can be observed that the catalyst particles are in very strict contact with the trapped soot, allowing many contact points between the soot and the catalyst, resulting in higher catalytic activity of the CDPF.

The Raman spectrum of a sample of the spent CDPF (Figure 5) evidenced the typical peaks of the SiC carrier at 295, 784 and 964 cm^{-1} [49]. Moreover, in Figure 5 the peaks relevant to soot are also present, and they have been attributed to bands G, D1, D3 and D4, as summarized in Table 1 [50].

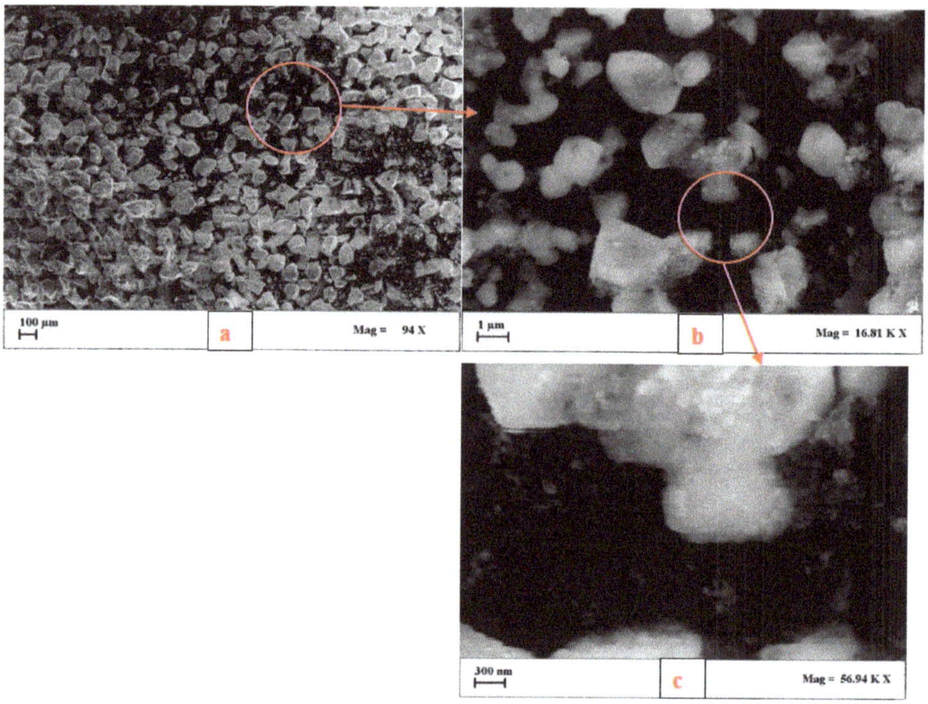

Figure 4. SEM images at different magnitudes (**a**), 94 X; (**b**), 16.81 X; (**c**), 56.94 X of the catalytic filter after the soot deposition stage and with a soot load of 5 g/L.

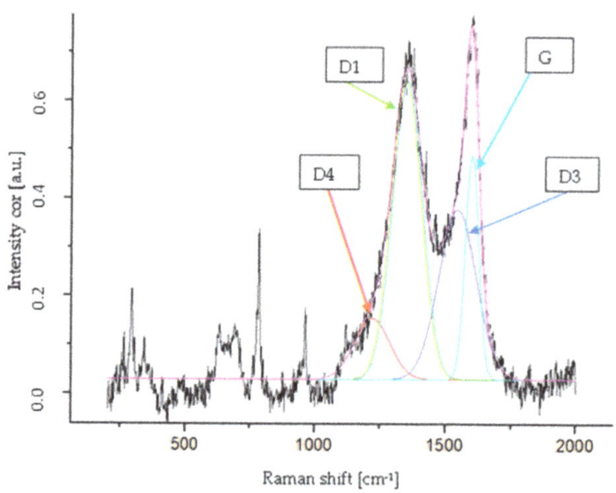

Figure 5. Fitting of the curve representing the Raman spectrum of diesel soot with band identification.

Table 1. Raman bands and vibration modes reported for soot.

Band	Raman Shift (cm^{-1})	Vibration Mode [50]
G	1602	Ideal graphitic lattice (E_{2g}-symmetry)
D1	1356	Disordered graphitic lattice (graphene layer edges, A_{1g} symmetry)
D3	1548	Amorphous carbon (Gaussian line shape)
D4	1224	Disordered graphitic lattice (A_{1g} symmetry)

3.3. Fresh Catalytic Samples

The XRD patterns of the prepared copper ferrite in comparison with cubic (database 77-0010), tetragonal (database 34-0425) and commercial (Merck KGaA, Darmstadt, Germany) forms are reported in Figure S3. XRD analysis shows the presence in the prepared catalyst of the typical peaks of tetragonal and cubic forms of $CuFe_2O_4$, whose average crystallite dimensions, calculated using the Scherrer formula, are about 20 nm. Two minor peaks ascribed to low amounts of oxides (CuO and Fe_2O_3) are also present.

The optimized procedure for the deposition of the active species, coupled to the modified porosity of the SiC monoliths, resulted in a uniform and homogeneous distribution of the copper ferrite not only on the monolith walls, but also in the porosities. The comparison of the catalytic and the bare samples in terms of BET specific surface area evidenced that the deposition of the active species resulted in a decrease in this parameter from 2.20 to 0.40 m^2/g. This result could be ascribed to the catalyst placement and the inside roughness caused by acid treatment. At this point, it is it is important to emphasize that this cannot be considered a negative result. In fact, these catalytic systems properly developed for the limitation of soot emissions are specifically applied to heterogeneous solid–solid–gas systems, in which the contact and the interactivity between the solid matrices (soot and catalyst) and the gas phase (mainly O_2) are the main parameters influencing the feasibility of soot combustion. Therefore, the maximization of these features is mandatory. In the catalytic systems investigated in this work, an intensified soot–catalyst contact was realized, as reported in our previous works [44], highlighting the very good catalytic activity towards soot oxidation with the O_2 present in the engine exhausts.

The SEM images shown in Figure 6 evidenced that the active species were homogeneously distributed, confirming previously reported results [23,44]. Moreover, no cracks formed due to preliminary acid treatment (Figure 6a), and the pores were not plugged after the catalyst deposition (Figure 6b). A deeper analysis of Figure 6 provides evidence that the catalyst aggregates formed a compact porous layer on the SiC granules (Figure 6c), which can be optimal for enhancing the soot–catalyst contact during the regeneration stage. Moreover, the catalyst deposition procedure allowed the deposition of the active species also in the pores of the DPF channel walls (Figure 6d), without their plugging.

The SEM-EDX analysis enabled the mapping of the elements present on the filter (Figure S4), and the detected elements were the structural ones (C, O and Si) and the catalyst active species (Cu and Fe).

The ultrasound adherence tests, shown in Figure S5 as weight loss (%) vs. number of cycles, evidenced weight losses much lower than those reported in the literature for washcoated supports [51]. These results proved that the preliminary calcination at 1000 °C for 48 h favored the adhesion of the active species on the SiC granules even in the absence of a washcoat, due to the formation of SiO_2 streaks on the SiC granules, as previously reported [52].

The H_2-TPR profile is shown in Figure 7 as H_2 consumption vs. temperature.

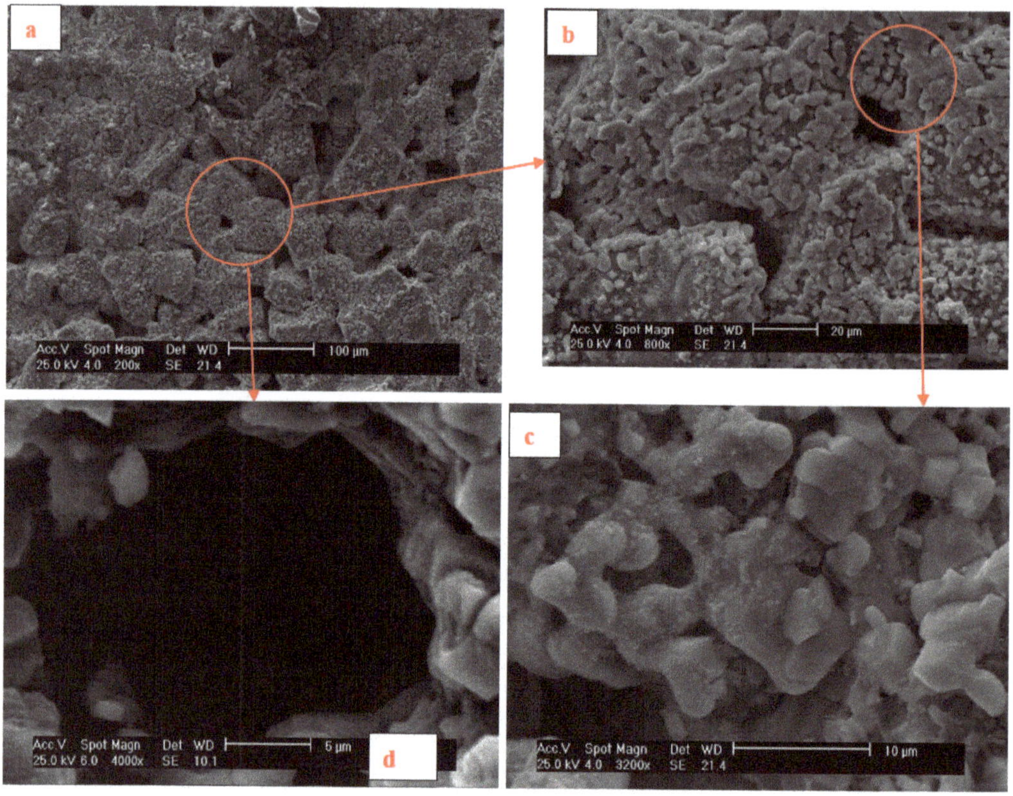

Figure 6. SEM images of the 30 %wt $CuFe_2O_4$ loaded monolith at various magnitudes: (**a**) 190 KX, (**b**) 965 KX, (**c**) 10.00 KX and (**d**) 25.09 KX.

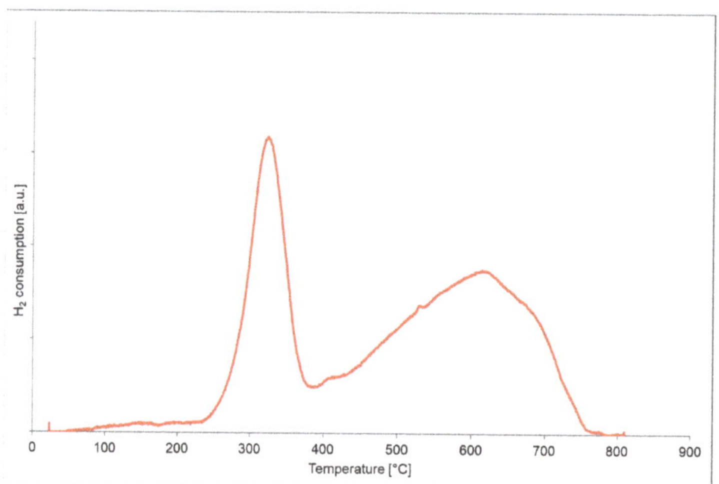

Figure 7. H_2-TPR profile of a SiC monolith loaded with 30 %wt of $CuFe_2O_4$.

Two pronounced reduction peaks, at about 300 °C and 610 °C, were present, which were attributed to the reduction of copper ferrite at lower temperatures and of Fe_3O_4 to Fe at higher temperatures [53].

By considering the two reduction reactions, it is possible to calculate the total amount of H_2 consumed for Cu moles (H_2/Cu ratio). In this case, this value was about 4, which:

i. Is consistent with the amount necessary to obtain the complete reduction of $CuFe_2O_4$ to Cu and Fe following the overall reaction (Equation (1)); and
ii. Corresponds to a load of $CuFe_2O_4$ equal to about 30.01 %wt, which is in very good agreement with the estimated 30 %wt of $CuFe_2O_4$ on the catalytic sample.

$$CuFe_2O_4 + 4H_2 \rightarrow Cu + 2Fe + 4H_2O \quad (1)$$

3.4. Experimental Tests in the Engine Test Cell

Both the soot loading and regeneration stages were performed under properly selected operating conditions in order to consider the influence of the exhaust flow rate on soot loading in terms of soot flow rate and soot cake properties. In particular, steady-state conditions were used to perform the soot loading, with a standard injection strategy consisting of pilot, pre- and main injections. In this phase, a constant value for the exhaust flow rate was used, aiming at monitoring the effective soot accumulation. In fact, two main terms are responsible for the pressure drop across the DPF:

i. ΔP_{filter}, due to the presence of the clean filter in the exhaust pipe, depending on both gas flow parameters (volumetric flow rate, viscosity and density) and filter geometry (volume, filter cell size, wall thickness and filter length); and
ii. ΔP_{soot}, related to the soot accumulation in the DPF, mainly in terms of soot cake properties [54].

During the filtration, the soot mass flow rate was calculated using the following formula:

$$\dot{m}_{soot} = \frac{\dot{m}_{exh}}{\rho} * C_{soot} \quad (2)$$

where $\dot{m}_{exh}$ and ρ are the mass flow rate (in kg/h) and density (in mg/m^3) of the exhaust gas at the inlet of the DPF, while C_{soot} is the PM concentration in mg/m^3 measured by the smoke meter. During the experimental tests, the soot concentration across the filters was continuously monitored in order to estimate the DPF efficiency, in terms of PM, using the following equation:

$$\eta_{filtration} = \frac{C_{soot\_in} - C_{soot\_out}}{C_{soot\_in}} * 100 \quad (3)$$

where C_{soot_in} and C_{soot_out} are the PM concentrations in mg/m^3 upstream and downstream of the filters, respectively.

The measurement of torque, fuel consumption, exhaust flow rate, T and DP across the filter was carried out to compare the engine performance during the filtration stage. The PSD measurements were carried out upstream and downstream of the CDPF. The size-dependent filter removal efficiency during the accumulation was evaluated for each diameter using the following equation:

$$\eta_{filtration} = \frac{(Nparticles)_{in} - (Nparticles)_{out}}{(Nparticles)_{in}} * 100 \quad (4)$$

where the subscripts "*in*" and "*out*" indicate the number of particles before and after the filtration, respectively. Moreover, the PN emission evolution during the active regeneration was detected. This represents a critical issue in a future perspective; in fact, the regeneration procedure could be entailed in the type of approval cycle (i.e., RDE).

The experimental tests during the soot loading phase were performed under four different engine loads in the range of 22–36%, with an engine speed equal to 2140 rpm and standard ECU calibration. The main engine variables are reported in Table S4. Engine load conditions were selected to compare filter performance also when the passive regeneration occurred with the CDPF (e.g., loads equal to 33% and 36%).

Regarding the active regeneration, PSD was correlated with the time history of DP across the CDPF and inlet/outlet temperatures. According to the description reported in a previous work [43], the procedure was actuated when a soot loading of 5 g/L had been reached: this quantity, not far from that reported in the literature [55,56], was recommended by the filter manufacturer as a value that does not impact on engine performance. The comparison between the CDPF and the bare DPF is shown in Figure 8 in terms of normalized pressure drop and temperature during the regeneration phase in the same engine operating conditions. The beneficial effect of the catalyst deposited on the DPF is evident: the DP curve decreases, meaning that soot oxidation occurrence is needed at temperatures lower than 400 °C for the CDPF (Figure 8a) and about 600 °C for the bare DPF (Figure 8b). Regarding the contents of the following sections, the whole regeneration process of the CDPF was divided into three consecutive phases (Figure 8a):

i. Start-up (0–200 s), in which any soot oxidation occurs;
ii. Soot combustion (200–450 s);
iii. End of regeneration, in which constant DP (after 450 s) and constant differences between temperatures across the CDPF are present.

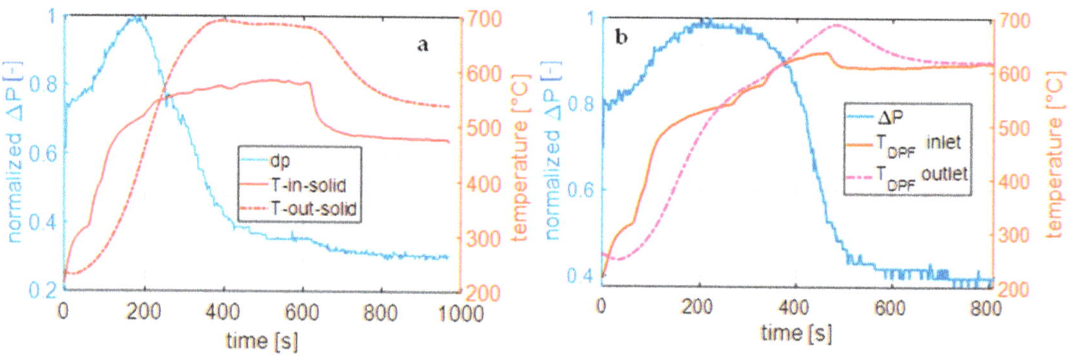

Figure 8. Trend of pressure drop and temperature across the (**a**) CDPF and (**b**) DPF during the regeneration.

Particle Emission Measurements

In this section, the PSDs at the CDPF outlet are presented and discussed; the results are compared with those obtained for the DPF [43] to emphasize the differences in terms of particle emissions. The plots shown in Figures 9–12 correspond to the raw experimental PSD data measured in the engine tailpipe. Due to the complex setup adopted for the measurements and the intrinsic oscillations due to the typical wave propagation of the engine exhaust gas flow to be analyzed, the PSD data were expected to be noisy. However, in this case no digital filter to generate a smoother data distribution was applied to render this aspect for readers and better evaluate the quality of the data and the difficulty in obtaining them. Regarding the particle size determination, Figures 9–12 show the mobility diameter (Dm) in a log scale; therefore, the range of values mentioned in the text can be considered correct despite the data oscillation trend. Moreover, the PSDs are represented on the Y-axis in Figures 9, 10 and 12 as dN/dln (Dm), where N is the number density and Dm is the mobility diameter.

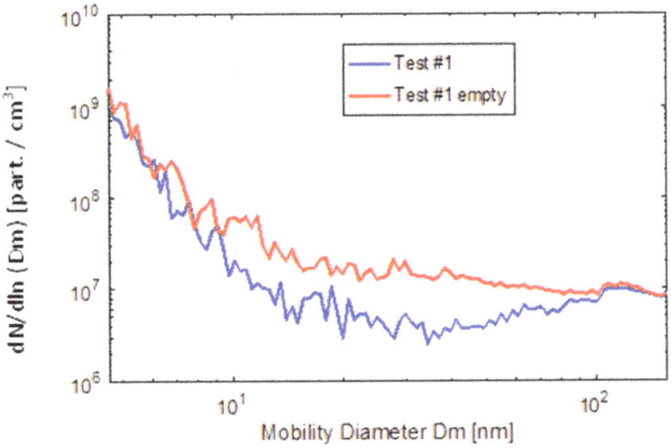

Figure 9. PSDs at the CDPF outlet for test case #1 during the accumulation (blue line) and just after the active regeneration ended (red line).

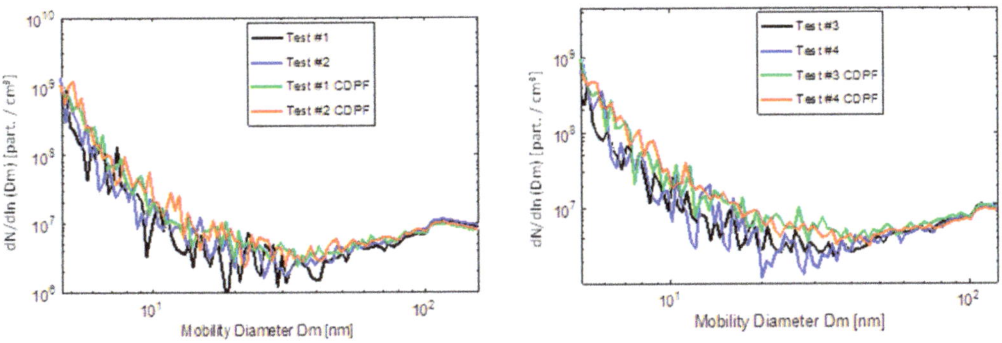

Figure 10. Comparison of outlet PSDs during accumulation for test cases #1 and #2 (**left side**) and for test cases #3 and #4 (**right side**).

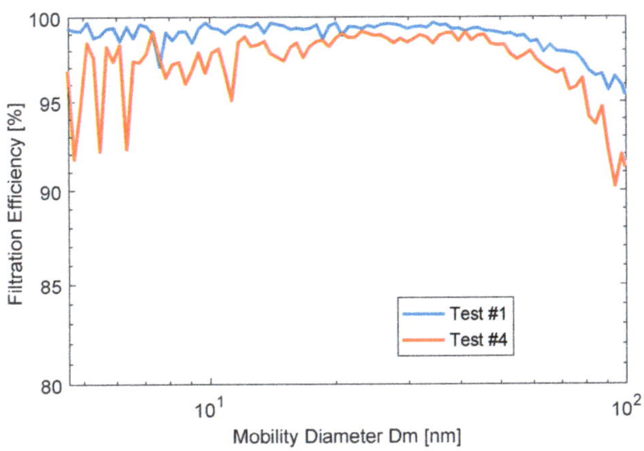

Figure 11. PN removal efficiency of the CDPF for test cases #1 and #4.

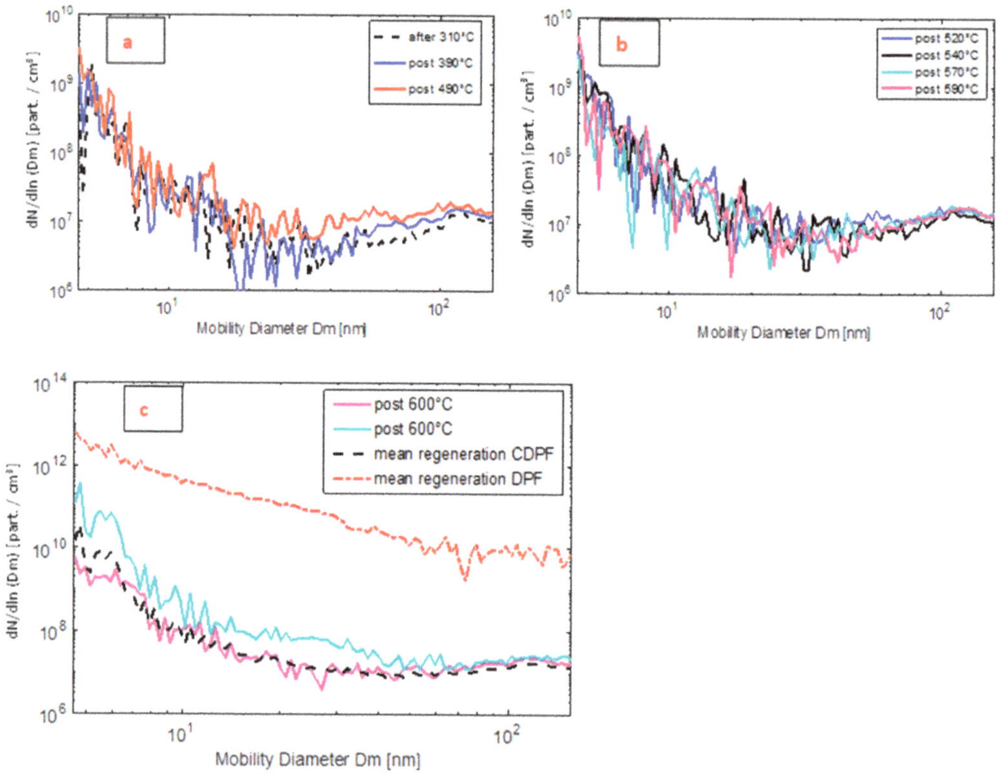

Figure 12. PSDs during the evolution of the regeneration at the CDPF outlet, at the start-up of the regeneration (**a**), in the range of 200–450 s (**b**) and at the end (**c**).

Figure 9 shows the PN concentration at the CDPF outlet for test case #1. The PSD bimodality [43] was still observed, and, after a minimum around 20–30 nm, the particle number increased for larger diameters (blue line). Differently from what was recently measured by authors with a regular DPF in the same conditions [43], a remarkable difference between the empty and partially filled filter was observed; in particular, an appreciable increase in the range of 10–100 nm was clearly visible (red line). The latter is due to the high reactivity of the CPDF, which ensures complete filter emptying during the active regeneration. However, the slight PN increase disappears after a few minutes of accumulation.

Figure 10 shows a comparison of the catalytic and the bare filters. In Figure 10a, the PSDs for test cases #1 and #2 (low engine speed) are reported: the two filters exhibited very similar results in terms of both shape and concentration across the whole range investigated. On the other hand, in Figure 10b (test cases #3 and #4—high engine speed) the CDPF presents a slight increase in PN in both test cases. Focusing on test case #3, the peak of 5 nm particles is three times higher than that observed for the DPF (green and black lines, respectively). The trend was observed up to 40 nm, after which the difference between the PSDs measured with the two filters was negligible. Similar remarks could be made with reference to the comparison of the PSDs obtained in test case #4. The increase in the number of 5 nm particles confirmed that, in the CDPF, the passive regeneration occurs starting from 33% of engine load [21]. In fact, while for the standard filter the PN removal efficiency was comparable, for the CDPF at 36% of engine load a slight reduction was detected. This results in partial filter emptying, thus allowing a reduction in the frequency of active regeneration. It is worth remarking that the particle emissions detected were much lower than those achieved during active regeneration; therefore, a reduction in both pollutants and fuel consumption can be achieved by operating the CDPF rather than the DPF.

The above results were also confirmed by the filtration efficiency estimation, as defined by Equation (3), shown in Figure 11.

In accordance with the DPF results, the highest PN removal was observed for test case #1, due to the lowest spatial velocity. In particular, the CDPF seemed to be very effective also for the sub-23 nm particles and exhibited a constant efficiency, below 40 nm, close to 99%. The narrow difference, compared with the DPF [43], is due to the lower mean pore diameter which affects the particle removal efficiency. On the other hand, in test case #4 the passive regeneration event leads to a slight decrease in efficiency for particle diameters below 10 nm. This phenomenon was not noticed for the standard DPF, which exhibited very close values for all the investigated conditions. For particles larger than 60 nm, the filtration efficiency began to decrease, and the CDPF exhibited similar behavior to the DPF.

Figure 12a shows the PSDs measured at the CDPF exit during the start-up of the regeneration. The dashed black line represents the PN distribution obtained by actuating the fuel after injection. The concentrations of particles with sizes below 10 nm were one order of magnitude lower than in the DPF and remained almost constant for the whole particle size range investigated; the PSDs are like those measured during the accumulation phase, as proof of the high reactivity of the catalyst. The benefits of the CDPF are even more striking comparing the PSDs when the fuel post-injection is actuated (blue and red lines). In fact, no significant increase in particle emissions was observed; moreover, the results highlight the absence of a correlation between the concentration peak of particles with sizes below 10 nm and the post-injection fuel quantity. In particular, the particle emissions were three orders of magnitude lower than the standard filter and remained two orders of magnitude lower for particle sizes larger than 50 nm. The slight increase in the concentration of particles in the range of 30–60 nm (red line) was probably due to the start of the soot oxidation, which resulted in the soot layer fragmentation responsible for a lower filtration efficiency towards the particles in this range.

Figure 12b shows the PSDs measured at the CDPF exit during the second phase of the regeneration process (200–450 s). The plots refer to different mean inlet temperatures to analyze the evolution of PSDs in the second stage of regeneration. Compared to the results obtained with the standard filter, the CDPF exhibited a bimodality of the PSDs also observed during the accumulation phase, with a peak of 5 nm particles equal to 3×10^9 #/cm^3, three orders of magnitude lower than for the DPF. Very similar values were detected during this stage, which emphasizes the improved soot oxidation thanks to the catalytic activity. The reduction in PN concentration at the outlet of the CDPF was confirmed by the sharp increase in the outlet temperature: while for the standard filter it exceeded the inlet temperature after 380 s [43], for the catalytic filter, this occurred much earlier, after 220 s. Moreover, the temperature increase across the filter reached 100 °C, two times higher than that detected across the DPF. Given these considerations, it can be asserted that the local oxidation temperatures were much higher and that complete combustion of the particles was achieved.

Figure 12c shows the PSDs at the end of the regeneration, identified by the quasi-constant pressure drop across the CDPF and the decrease in the outlet temperature. The magenta and cyan lines refer to two consecutive measurements collected in a time window of one minute in the same inlet conditions. A PN peak of 2×10^{11} was detected for particles with sizes below 10 nm in the later measurement (cyan line), which is more than one order of magnitude higher than the earlier measurement (red line). Furthermore, an appreciable increase in the range of 20–40 nm was noticed, which yielded a PSD shape not observed in the earlier measurements. This behavior can be explained considering that, despite the same inlet temperature being used, the DPF core exhibited a higher temperature in the former case, thus promoting the oxidation of the particles with lower diameters. Furthermore, this result confirms the complete emptying of the CDPF, previously evidenced in Figure 9.

The dashed black line in Figure 12 shows the mean PSD during the whole regeneration process for the CDPF. The comparison with the corresponding values detected for the DPF, represented by the dashed red line, highlights the impressive reduction in particle emissions achieved by the CDPF in the whole range investigated. In particular, the peak of the 5 nm PN dropped from 4×10^{12} to 3×10^{10}, two orders of magnitude lower, while for the particle size above 50 nm the reduction reached three orders of magnitude, from 10^{10} to 10^{7}.

This very important result highlighted that the use of a well-designed CDPF can allow the oxidation of accumulated soot at lower temperatures and decrease in particle emissions during the regeneration phase, not only with respect to a bare DPF, but also with respect to recent studies in the literature, in which increases in emissions of particles of sizes above and below 23 nm were observed [45,57].

4. Conclusions

In this work, starting from the results previously obtained on soot oxidation during DPF regeneration [21–23], the research focused on the detailed PSD analysis of particles emitted during the active regeneration of a $CuFe_2O_4$ catalyzed DPF at the exhaust of an L-D common-rail diesel engine. Different tests were performed under different engine operating conditions, aiming at estimating the PN removal efficiency of the CDPF in the range of 5–100 nm. The filtration efficiency in the investigated PN range was evaluated by alternately sampling the particles upstream and downstream of the CDPF. In particular, the dynamics of the PN size distributions during the regeneration process were investigated. The results of the experimental tests evidenced that, during the start-up of the regeneration, differently from the standard DPF, no significant increase in particle emissions at the CDPF outlet was observed. In fact, particle emissions three orders of magnitude lower than those detected when regenerating the standard filter were observed, and they remained two orders of magnitude lower for particle sizes larger than 50 nm. With the regeneration going on, in the range of 200–450 s, the PSDs measured at the CDPF outlet exhibited the bimodality observed during the accumulation phase, with a peak that was three orders of magnitude lower than that detected in the case of the bare DPF. At the end of the regeneration, characterized by a quasi-constant pressure drop across the CDPF, an impressive reduction in PN emissions was still observed vs. the standard DPF, with PN values that were two and three orders of magnitude lower for the particle sizes of 5 nm and above 50 nm, respectively. This very important result highlighted that the use of a well-designed CDPF can allow the oxidation of the accumulated soot at lower temperatures and a decrease in particle emissions during the regeneration phase, not only with respect to a bare DPF, but also with respect to recent studies in the literature, in which increases in emissions of particles above and below 23 nm were observed.

Supplementary Materials: The following supporting information can be downloaded at: https://www.mdpi.com/article/en16104071/s1, Table S1: Geometrical data of the bare DPF; Table S2: Engine technical data; Table S3: Sensors accuracy; Table S4: Engine conditions during accumulation phase in 4 test cases; Figure S1: SEM image of the soot emitted by the Diesel engine used in the experimental tests; Figure S2: SEM image (left) and EDX mapping (right) of soot emitted by the Diesel engine as trapped by the CDPF; Figure S3: XRD pattern of the prepared $CuFe_2O_4$ compared with cubic, tetragonal and commercial copper ferrite; Figure S4: SEM image and distribution of elements, obtained by EDX element mapping, for the 30%wt $CuFe_2O_4$ loaded DPF; Figure S5: Ultrasonic tests performed on 30 %wt $CuFe_2O_4$ catalysed monolith.

Author Contributions: Conceptualization, E.M., B.R., G.D.F., M.S., I.A. and V.P.; methodology, E.M., B.R., G.D.F., M.S., I.A. and V.P.; software, E.M., B.R., G.D.F., M.S., I.A. and V.P.; validation, E.M., B.R., G.D.F., M.S., I.A. and V.P.; formal analysis, E.M., B.R., G.D.F., M.S., I.A. and V.P.; investigation, E.M., B.R., G.D.F., M.S., I.A. and V.P.; resources, E.M., B.R., G.D.F., M.S., I.A. and V.P.; data curation, E.M., B.R., G.D.F., M.S., I.A. and V.P.; writing—original draft preparation, E.M., B.R., G.D.F., M.S., I.A., V.P.; writing—review and editing, E.M., B.R., G.D.F., M.S., I.A. and V.P.; visualization, E.M., B.R., G.D.F., M.S., I.A. and V.P.; supervision, I.A. and V.P.; project administration, I.A. and V.P. All authors have read and agreed to the published version of the manuscript.

Funding: This research received no external funding.

Data Availability Statement: Not available.

Conflicts of Interest: The authors declare no conflict of interest.

References

1. Beatrice, C.; Di Iorio, S.; Guido, C.; Napolitano, P. Detailed characterization of particulate emissions of an automotive catalyzed DPF using actual regeneration strategies. *Exp. Therm. Fluid Sci.* **2012**, *39*, 45–53. [CrossRef]
2. Alexandrino, K. Comprehensive Review of the Impact of 2,5-Dimethylfuran and 2-Methylfuran on Soot Emissions: Experiments in Diesel Engines and at Laboratory-Scale. *Energy Fuels* **2020**, *34*, 6598–6623. [CrossRef]
3. Zhang, Y.; Xia, C.; Liu, D.; Zhu, Y.; Feng, Y. Experimental investigation of the high-pressure SCR reactor impact on a marine two-stroke diesel engine. *Fuel* **2023**, *335*, 127064. [CrossRef]
4. Pacino, A.; La Porta, C.; La Rocca, A.; Cairns, A. Copper leaching effects on combustion characteristics and particulate emissions of a direct injection high pressure common rail diesel engine. *Fuel* **2023**, *340*, 127536. [CrossRef]
5. Apicella, B.; Mancaruso, E.; Russo, C.; Tregrossi, A.; Oliano, M.M.; Ciajolo, A.; Vaglieco, B.M. Effect of after-treatment systems on particulate matter emissions in diesel engine exhaust. *Exp. Therm. Fluid Sci.* **2020**, *116*, 110107. [CrossRef]
6. Bagi, S.; Kamp, C.J.; Sharma, V.; Aswath, P.B. Multiscale characterization of exhaust and crankcase soot extracted from heavy-duty diesel engine and implications for DPF ash. *Fuel* **2020**, *282*, 118878. [CrossRef]
7. Meloni, E.; Palma, V. Most Recent Advances in Diesel Engine Catalytic Soot Abatement: Structured Catalysts and Alternative Approaches. *Catalysts* **2020**, *10*, 745. [CrossRef]
8. Gren, L.; Malmborg, V.B.; Jacobsen, N.R.; Shukla, P.C.; Bendtsen, K.M.; Eriksson, A.C.; Essig, Y.J.; Krais, A.M.; Loeschner, K.; Shamun, S.; et al. Effect of Renewable Fuels and Intake O_2 Concentration on Diesel Engine Emission Characteristics and Reactive Oxygen Species (ROS) Formation. *Atmosphere* **2020**, *11*, 641. [CrossRef]
9. Fang, J.; Meng, Z.; Li, J.; Du, Y.; Qin, Y.; Jiang, Y.; Bai, W.; Chase, G.G. The effect of operating parameters on regeneration characteristics and particulate emission characteristics of diesel particulate filters. *Appl. Therm. Eng.* **2019**, *148*, 860–867. [CrossRef]
10. Meloni, E.; Palma, V. Low Temperature Microwave Regeneration of Catalytic Diesel Particulate Filter. *Chem. Eng. Trans.* **2018**, *70*, 721–726. [CrossRef]
11. Saffaripour, M.; Chan, T.W.; Liu, F.; Thomson, K.A.; Smallwood, G.J.; Kubsh, J.; Brezny, R. Effect of Drive Cycle and Gasoline Particulate Filter on the Size and Morphology of Soot Particles Emitted from a Gasoline-Direct-Injection Vehicle. *Environ. Sci. Technol.* **2015**, *49*, 11950–11958. [CrossRef] [PubMed]
12. Kong, X.; Li, Z.; Shen, B.; Wu, Y.; Zhang, Y.; Cai, D. Simulation of flow and soot particle distribution in wall-flow DPF based on lattice Boltzmann method. *Chem. Eng. Sci.* **2019**, *202*, 169–185. [CrossRef]
13. Yamamoto, K.; Nakamura, M.; Yane, H.; Yamashita, H. Simulation on catalytic reaction in diesel particulate filter. *Catal. Today* **2010**, *153*, 118–124. [CrossRef]
14. Yamamoto, K.; Oohori, S.; Yamashita, H.; Daido, S. Simulation on soot deposition and combustion in diesel particulate filter. *Proc. Combust. Inst.* **2009**, *32*, 1965–1972. [CrossRef]
15. Oi-Uchisawa, J.; Obuchi, A.; Wang, S.; Nanba, T.; Ohi, A. Catalytic performance of Pt/MOx loaded over SiC-DPF for soot oxidation. *Appl. Catal. B Environ.* **2003**, *43*, 117–129. [CrossRef]
16. Zhang, M.; Ge, Y.; Wang, X.; Peng, Z.; Tan, J.; Hao, L.; Lv, L.; Wang, C. An investigation into the impact of burning diesel/lubricant oil mixtures on the nature of particulate emissions: Implications for DPF ash-loading acceleration method. *J. Energy Inst.* **2020**, *93*, 1207–1215. [CrossRef]
17. Giechaskiel, B. Particle Number Emissions of a Diesel Vehicle during and between Regeneration Events. *Catalysts* **2020**, *10*, 587. [CrossRef]
18. Jiaqiang, E.; Zhao, M.; Zuo, Q.; Zhang, B.; Zhang, Z.; Peng, Q.; Han, D.; Zhao, X.; Deng, Y. Effects analysis on diesel soot continuous regeneration performance of a rotary microwave-assisted regeneration diesel particulate filter. *Fuel* **2020**, *260*, 116353. [CrossRef]
19. Kurien, C.; Srivastava, A.K.; Lesbats, S. Experimental and computational study on the microwave energy based regeneration in diesel particulate filter for exhaust emission control. *J. Energy Inst.* **2020**, *93*, 2133–2147. [CrossRef]
20. Kurien, C.; Srivastava, A.K.; Gandigudi, N.; Anand, K. Soot deposition effects and microwave regeneration modelling of diesel particulate filtration system. *J. Energy Inst.* **2020**, *93*, 463–473. [CrossRef]
21. Rossomando, B.; Arsie, I.; Meloni, E.; Palma, V.; Pianese, C. *Experimental Test on the Feasibility of Passive Regeneration in a Catalytic DPF at the Exhaust of a Light-Duty Diesel Engine*; SAE Technical Paper 2019-24-0045; SAE International (Europe): London, UK, 2019.
22. Rossomando, B.; Arsie, I.; Meloni, E.; Palma, V.; Pianese, C. *Experimental Testing of a Low Temperature Regenerating Catalytic DPF at the Exhaust of a Light-Duty Diesel Engine*; SAE Technical Paper 2018-01-0351; SAE International (Europe): London, UK, 2018.
23. Meloni, E.; Palma, V.; Vaiano, V. Optimized microwave susceptible catalytic diesel soot trap. *Fuel* **2017**, *205*, 142–152. [CrossRef]
24. Leach, F.C.; Davy, M.; Terry, B. Combustion and emissions from cerium oxide nanoparticle dosed diesel fuel in a high speed diesel research engine under low temperature combustion (LTC) conditions. *Fuel* **2021**, *288*, 119636. [CrossRef]

25. Zhang, Z.-H.; Balasubramanian, R. Effects of Cerium Oxide and Ferrocene Nanoparticles Addition as Fuel-Borne Catalysts on Diesel Engine Particulate Emissions: Environmental and Health Implications. *Environ. Sci. Technol.* **2017**, *51*, 4248–4258. [CrossRef] [PubMed]
26. He, J.; Yao, P.; Qiu, J.; Zhang, H.; Jiao, Y.; Wang, J.; Chen, Y. Enhancement effect of oxygen mobility over $Ce_{0.5}Zr_{0.5}O_2$ catalysts doped by multivalent metal oxides for soot combustion. *Fuel* **2021**, *286*, 119359. [CrossRef]
27. Andana, T.; Piumetti, M.; Bensaid, S.; Russo, N.; Fino, D. Heterogeneous mechanism of NO_x-assisted soot oxidation in the passive regeneration of a bench-scale diesel particulate filter catalyzed with nanostructured equimolar ceria-praseodymia. *Appl. Catal. A Gen.* **2019**, *583*, 117136. [CrossRef]
28. Mori, K.; Jida, H.; Kuwahara, Y.; Yamashita, H. CoO_x-decorated CeO_2 heterostructures: Effects of morphology on their catalytic properties in diesel soot combustion. *Nanoscale* **2020**, *12*, 1779–1789. [CrossRef]
29. Serhan, N.; Tsolakis, A.; Wahbi, A.; Martos, F.; Golunski, S. Modifying catalytically the soot morphology and nanostructure in diesel exhaust: Influence of silver De-NOx catalyst (Ag/Al_2O_3). *Appl. Catal. B Environ.* **2019**, *241*, 471–482. [CrossRef]
30. Ji, F.; Men, Y.; Wang, J.; Sun, Y.; Wang, Z.; Zhao, B.; Tao, X.; Xu, G. Promoting diesel soot combustion efficiency by tailoring the shapes and crystal facets of nanoscale Mn_3O_4. *Appl. Catal. B Environ.* **2019**, *242*, 227–237. [CrossRef]
31. Singer, C.; Kureti, S. Soot oxidation in diesel exhaust on manganese oxide catalyst prepared by flame spray pyrolysis. *Appl. Catal. B Environ.* **2020**, *272*, 118961. [CrossRef]
32. Kuwahara, Y.; Fujibayashi, A.; Uehara, H.; Mori, K.; Yamashita, H. Catalytic combustion of diesel soot over Fe and Ag-doped manganese oxides: Role of heteroatoms in the catalytic performances. *Catal. Sci. Technol.* **2018**, *8*, 1905–1914. [CrossRef]
33. Di Sarli, V.; Landi, G.; Lisi, L.; Saliva, A.; Di Benedetto, A. Catalytic diesel particulate filters with highly dispersed ceria: Effect of the soot-catalyst contact on the regeneration performance. *Appl. Catal. B Environ.* **2016**, *197*, 116–124. [CrossRef]
34. Jeftić, M.; Yu, S.; Han, X.; Reader, G.T.; Wang, M.; Zheng, M. Effects of Postinjection Application with Late Partially Premixed Combustion on Power Production and Diesel Exhaust Gas Conditioning. *J. Combust.* **2011**, *2011*, 891096. [CrossRef]
35. D'Aniello, F.; Rossomando, B.; Arsie, I.; Pianese, C. *Development and Experimental Validation of a Control Oriented Model of a Catalytic DPF*; SAE Technical Paper 2019-01-0985; SAE International (Europe): London, UK, 2019.
36. Lao, C.T.; Akroyd, J.; Eaves, N.; Smith, A.; Morgan, N.; Bhave, A.; Kraft, M. Modelling particle mass and particle number emissions during the active regeneration of diesel particulate filters. *Proc. Combust. Inst.* **2019**, *37*, 4831–4838. [CrossRef]
37. Wang, D.-Y.; Tan, P.-Q.; Zhu, L.; Wang, Y.-H.; Hu, Z.-Y.; Lou, D.-M. Novel soot loading prediction model of diesel particulate filter based on collection mechanism and equivalent permeability. *Fuel* **2021**, *286*, 119409. [CrossRef]
38. Dwyer, H.; Ayala, A.; Zhang, S.; Collins, J.; Huai, T.; Herner, J.; Chau, W. Emissions from a diesel car during regeneration of an active diesel particulate filter. *J. Aerosol Sci.* **2010**, *41*, 541–552. [CrossRef]
39. Meng, Z.; Chen, C.; Li, J.; Fang, J.; Tan, J.; Qin, Y.; Jiang, Y.; Qin, Z.; Bai, W.; Liang, K. Particle emission characteristics of DPF regeneration from DPF regeneration bench and diesel engine bench measurements. *Fuel* **2020**, *262*, 116589. [CrossRef]
40. Wihersaari, H.; Pirjola, L.; Karjalainen, P.; Saukko, E.; Kuuluvainen, H.; Kulmala, K.; Keskinen, J.; Rönkkö, T. Particulate emissions of a modern diesel passenger car under laboratory and real-world transient driving conditions. *Environ. Pollut.* **2020**, *265*, 114948. [CrossRef]
41. Liati, A.; Schreiber, D.; Dasilva, Y.A.R.; Eggenschwiler, P.D. Ultrafine particle emissions from modern Gasoline and Diesel vehicles: An electron microscopic perspective. *Environ. Pollut.* **2018**, *239*, 661–669. [CrossRef]
42. Meng, Z.; Li, J.; Fang, J.; Tan, J.; Qin, Y.; Jiang, Y.; Qin, Z.; Bai, W.; Liang, K. Experimental study on regeneration performance and particle emission characteristics of DPF with different inlet transition sections lengths. *Fuel* **2020**, *262*, 116487. [CrossRef]
43. Rossomando, B.; Meloni, E.; De Falco, G.; Sirignano, M.; Arsie, I.; Palma, V. Experimental characterization of ultrafine particle emissions from a light-duty diesel engine equipped with a standard DPF. *Proc. Combust. Inst.* **2021**, *38*, 5695–5702. [CrossRef]
44. Palma, V.; Ciambelli, P.; Meloni, E.; Sin, A. Optimal $CuFe_2O_4$ load for MW susceptible catalysed DPF. *Chem. Eng. Trans.* **2013**, *35*, 727–732. [CrossRef]
45. R'mili, B.; Boréave, A.; Meme, A.; Vernoux, P.; Leblanc, M.; Noël, L.; Raux, S.; D'anna, B. Physico-Chemical Characterization of Fine and Ultrafine Particles Emitted during Diesel Particulate Filter Active Regeneration of Euro5 Diesel Vehicles. *Environ. Sci. Technol.* **2018**, *52*, 3312–3319. [CrossRef] [PubMed]
46. Salenbauch, S.; Sirignano, M.; Pollack, M.; D'anna, A.; Hasse, C. Detailed modeling of soot particle formation and comparison to optical diagnostics and size distribution measurements in premixed flames using a method of moments. *Fuel* **2018**, *222*, 287–293. [CrossRef]
47. Sgro, L.A.; Sementa, P.; Vaglieco, B.M.; Rusciano, G.; D'anna, A.; Minutolo, P. Investigating the origin of nuclei particles in GDI engine exhausts. *Combust. Flame* **2012**, *159*, 1687–1692. [CrossRef]
48. De Filippo, A.; Maricq, M.M. Diesel Nucleation Mode Particles: Semivolatile or Solid? *Environ. Sci. Technol.* **2008**, *42*, 7957–7962. [CrossRef]
49. Burton, J.C.; Sun, L.; Pophristic, M.; Lukacs, S.J.; Long, F.H.; Feng, Z.C.; Ferguson, I.T. Spatial characterization of doped SiC wafers by Raman spectroscopy. *J. Appl. Phys.* **1998**, *84*, 6268–6273. [CrossRef]
50. Sadezky, A.; Muckenhuber, H.; Grothe, H.; Niessner, R.; Pöschl, U. Raman microspectroscopy of soot and related carbonaceous materials: Spectral analysis and structural information. *Carbon* **2005**, *43*, 1731–1742. [CrossRef]
51. Giani, L.; Cristiani, C.; Groppi, G.; Tronconi, E. Washcoating methods for Pd/gamma-Al_2O_3 deposition on metallic foams. *Appl. Catal. B Environ.* **2006**, *62*, 121–131. [CrossRef]

52. Palma, V.; Ciambelli, P.; Meloni, E.; Sin, A. Catalytic DPF microwave assisted active regeneration. *Fuel* **2015**, *140*, 50–61. [CrossRef]
53. Kameoka, S.; Tanabe, T.; Tsai, A.P. Self-assembled porous nano-composite with high catalytic performance by reduction of tetragonal spinel $CuFe_2O_4$. *Appl. Catal. A Gen.* **2010**, *375*, 163–171. [CrossRef]
54. Choi, S.; Oh, K.-C.; Lee, C.-B. The effects of filter porosity and flow conditions on soot deposition/oxidation and pressure drop in particulate filters. *Energy* **2014**, *77*, 327–337. [CrossRef]
55. Haralampous, O.A.; Kandylas, I.P.; Koltsakis, G.C.; Samaras, Z.C. Diesel particulate filter pressure drop Part 1: Modelling and experimental validation. *Int. J. Engine Res.* **2004**, *5*, 149–162. [CrossRef]
56. Lapuerta, M.; Rodríguez-Fernández, J.; Oliva, F. Effect of soot accumulation in a diesel particle filter on the combustion process and gaseous emissions. *Energy* **2012**, *47*, 543–552. [CrossRef]
57. Yamada, H.; Inomata, S.; Tanimoto, H. Mechanisms of Increased Particle and VOC Emissions during DPF Active Regeneration and Practical Emissions Considering Regeneration. *Environ. Sci. Technol.* **2017**, *51*, 2914–2923. [CrossRef]

energies MDPI

Article

Experimental Characterization of Hydrocarbons and Nitrogen Oxides Production in a Heavy-Duty Diesel–Natural Gas Reactivity-Controlled Compression Ignition Engine

Giacomo Silvagni [1], Abhinandhan Narayanan [2], Vittorio Ravaglioli [1,*], Kalyan Kumar Srinivasan [2], Sundar Rajan Krishnan [2], Nik Collins [2], Paulius Puzinauskas [2] and Fabrizio Ponti [1]

1 Department of Industrial Engineering—DIN, University of Bologna, 40126 Bologna, Italy; giacomo.silvagni2@unibo.it (G.S.); fabrizio.ponti@unibo.it (F.P.)
2 Department of Mechanical Engineering, The University of Alabama, Tuscaloosa, AL 35487, USA; anarayanan2@crimson.ua.edu (A.N.); ksrinivasan@eng.ua.edu (K.K.S.); skrishnan@eng.ua.edu (S.R.K.); nscollins@crimson.ua.edu (N.C.); ppuzinauskas@eng.ua.edu (P.P.)
* Correspondence: vittorio.ravaglioli2@unibo.it

Abstract: Reactivity-Controlled Compression Ignition (RCCI) combustion is considered one of the most promising Low-Temperature Combustion (LTC) concepts aimed at reducing greenhouse gases for the transportation and power generation sectors. Due to the spontaneous combustion of a lean, nearly homogeneous mixture of air and low-reactivity fuel (LRF), ignited through the direct injection of a small quantity of high-reactivity fuel (HRF), RCCI (dual-fuel) shows higher efficiency and lower pollutants compared to conventional diesel combustion (CDC) if run at very advanced injection timing. Even though a HRF is used, the use of advanced injection timing leads to high ignition delays, compared to CDC, and generates high cycle-to-cycle variability, limited operating range, and high pressure rise rates at high loads. This work presents an experimental analysis performed on a heavy-duty single-cylinder compression ignited engine in dual-fuel diesel–natural gas mode. The objective of the present work is to investigate and highlight the correlations between combustion behavior and pollutant emissions, especially unburned hydrocarbons (HC) and oxides of nitrogen (NOx). Based on the analysis of crank-resolved pollutants measurements performed through fast FID and fast NOx systems under different engine operating conditions, two correlations were found demonstrating a good accordance between pollutant production and combustion behavior: Net Cyclic Hydrocarbon emission—cyclic IMEP variations (R^2 = 0.86), and Cyclic NOx—maximum value of the Rate of Heat Released (R^2 = 0.82).

Keywords: LTC; RCCI; CO_2 reduction; dual fuel; fast FID; fast NOx

Citation: Silvagni, G.; Narayanan, A.; Ravaglioli, V.; Srinivasan, K.K.; Krishnan, S.R.; Collins, N.; Puzinauskas, P.; Ponti, F. Experimental Characterization of Hydrocarbons and Nitrogen Oxides Production in a Heavy-Duty Diesel–Natural Gas Reactivity-Controlled Compression Ignition Engine. *Energies* **2023**, *16*, 5164. https://doi.org/10.3390/en16135164

Academic Editors: Tomasz Czakiert and Monika Kosowska-Golachowska

Received: 30 May 2023
Revised: 20 June 2023
Accepted: 3 July 2023
Published: 4 July 2023

1. Introduction

Over the recent years, thanks to a vast improvement in fast sensor and actuator technology and a profound understanding of the phenomena that affect the combustion process in IC engines, the feasibility of implementing cycle-by-cycle-based control to attain optimum engine performance and emissions is growing. With the push towards global decarbonization growing stronger, a combination of advanced combustion strategies and effective next-cycle control could be a pivotal pathway towards meeting the stringent emissions regulations for IC engines. Researchers have demonstrated the abilities of technologies like Homogeneous Charge Compression Ignition (HCCI) [1,2], Premixed Mixture Ignition in the End-gas Region (PREMIER) [3,4], pre-chamber combustion systems [5,6] and Reactivity-Controlled Compression Ignition (RCCI) [7,8] to produce high fuel conversion efficiencies while maintaining low engine-out emissions. Of these, the RCCI/dual-fuel Low-Temperature Combustion (LTC) strategy has proven to be a viable pathway to achieve ultra-low NOx emissions and good fuel conversion efficiencies (FCE) at both high and low load operations [9–16]. Hariharan et al. [17] experimentally studied the effect of various engine operating parameters like injection timing, intake pressure and injection pressure on emissions and performance of dual-fuel combustion with diesel–natural gas (NG) and polyoxymethylene di-methyl ether (POMDME)–NG as fuel combinations. At low loads (~5 bar IMEP), the authors were able to achieve indicated FCEs of >40% and indicated specific NOx values of ~0.1 g/kWh at injection timings ranging from 40–60 deg bTDC. However, these single-injection operating points were found to have high cyclic variations and high HC emissions. The authors were able to mitigate the negative impact of cyclic variations by adding a second injection closer to TDC and by operating at lower intake pressures. Cyclic variations, which arise due to fluctuations in local air fuel ratios and in-cylinder thermodynamics, have a profound negative impact on HC emissions and engine performance. Methods to quantify cyclic variations and to identify its cause have been extensively documented in the literature [18–23]. For example, Jha et al. [24] experimentally examined the effect of methane energy fraction at low loads in diesel–methane dual fuel combustion at early injection timings. They observed that increasing the energy fraction of methane from 50% to 90% at −50 deg aTDC start of injection decreased NOx emissions by a factor of 43; increased HC emissions by a factor of 9 and increased cyclic variations. The authors attributed the increase in cyclic variations to the decrease in reactivity of the mixture as methane energy fraction increases. Various diagnostics methods have been investigated to examine cyclic variations in dual fuel combustion. Cheng et al. [25] examined cyclic variations in diesel—methane dual fuel combustion using optical methods and used a proper orthogonal decomposition method to analyze and quantify cyclic variations. The increased fluctuations in the luminosity field were considered to be a marker for increased cyclic variations. They observed delayed onset of cyclic variations as the methane fraction increased at an injection timing of 15 deg bTDC.

The objective of the present work is to correlate the cause of high HC and NOx emissions to the combustion behavior in diesel–NG dual fuel combustion using cyclic emissions data obtained from fast pollutant measurement systems in combination with combustion data.

2. Experimental Setup

Experimental studies were performed on a single cylinder version of a heavy-duty PACCAR MX11 engine setup at the University of Alabama (UA). The engine specifications are provided in Table 1. The Single-Cylinder Research Engine (SCRE) was coupled to a 393 HP AC dynamometer and a Dyne Systems IL5 controller was used to control the engine speed, while engine load (as determined by the Indicated Mean Effective Pressure (IMEP)) was set by controlling fueling rates.

Table 1. Engine technical characteristics.

Engine Parameter	Values
Engine type	Single cylinder, four-stroke
Displaced volume	1.806 m^3
Maximum Torque	350 Nm @ 1000 rpm
Maximum Power	53 kW @ 1750 rpm
Injection System	Common Rail Injection system
Bore	123 mm
Stroke	152 mm
Compression ratio	18.5:1
Intake Valve	IVO: 350 deg aTDC, IVC: −150 deg aTDC
Exhaust Valve	EVO: 130 deg aTDC, EVC: −355 deg aTDC
Maximum Speed	2200 rpm
Maximum Pressure	245 bar

A common rail direct injection system was used to inject the High-Reactivity Fuel (HRF) and the rail pressure was controlled using a solenoid actuated inlet metering valve and rail pressure control valve. A National Instruments Direct-injector Control and Measurement device (DCM) interfaced with the Calibration Viewer software from Vieletech and was used to control injection timing and injection duration. The Low-Reactivity Fuel (LRF) was fumigated along with the intake air and the LRF flowrate was controlled using a Swagelok KLF series pressure regulator. The fuel flowrates were measured using Emerson Micromotion Coriolis flowmeters.

The intake air system uses an external air compressor which also conditions the high-pressure air (up to 5 bar) using a heatless desiccant dryer. The intake air mass flowrate was measured using a sonic nozzle, for which the pressure and temperature upstream of the nozzle were also measured to compute the air mass flowrate. The upstream pressure was measured using a Setra 206S pressure transducer and the temperature was measured using an Omega K-type thermocouple. A Kistler piezoelectric pressure transducer (6124A) coupled with a charge amplifier (5018) was used to measure the in-cylinder pressure, and the dynamic pressures of both the intake and exhaust were measured using Kistler piezoresistive pressure transducers. The in-cylinder pressure profile was phased using a BEI shaft encoder with 0.1 CAD resolution.

Cycle-resolved HC and NOx were obtained by installing CAMBUSTION HFR400 FID (Fast FID) and CAMBUSTION CLD500 (Fast NOx) measurement systems in the exhaust manifold, respectively (as close as possible to the exhaust port). By using the Fast FID and Fast NOx systems, which can sample the pollutants at 500 Hz, it was possible to obtain the "instantaneous" emissions for each engine cycle. As a result, since the output of such measurement systems were phased with the data acquisition system, it was possible to link the cylinder pressure trace during the combustion process and the respective pollutant production in a given cycle.

Moreover, to compare the present data with the literature, slow-speed (1 Hz) pollutant measurements were also performed by using a NOVA 5466 AK gas analyzer (present in a standard Richmond five-gas emission bench) and a CONTINENTAL UniNOx SNS14 sensor (100 Hz) mounted close to the Fast NOx probe. Figure 1 shows the UA SCRE test cell layout used for the present experimental efforts.

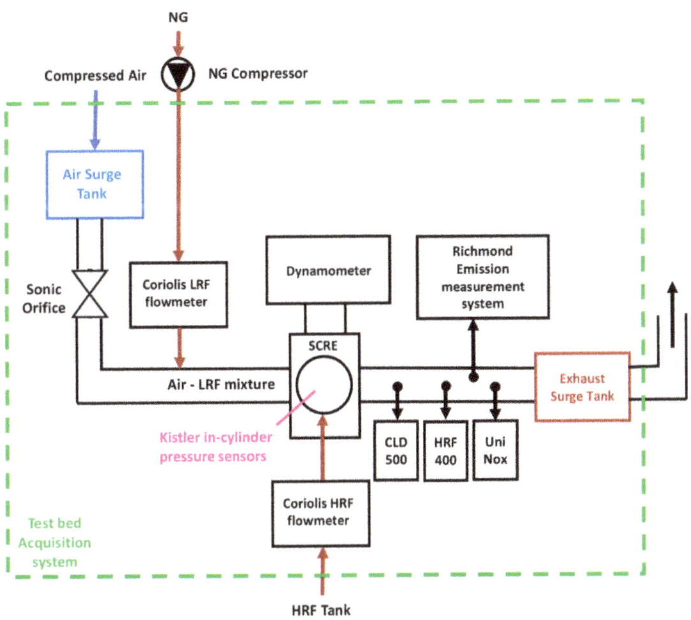

Figure 1. A schematic representation of the UA single-cylinder research engine setup used for the present experiments.

3. Experimental Methodology

As widely reported in the literature [26,27] and explained later in this paper, to achieve high combustion efficiencies with dual-fuel combustion, an early Start of Injection (SOI) is needed. With early SOIs, the shape of the apparent rate of heat release (RoHR) curve is "Gaussian" and it generates very low pollutants, especially NOx, and high indicated efficiency. However, dual-fuel combustion at very advanced SOIs leads to the ignition process (dictated by the HRF injection) becoming driven more by in-cylinder chemical reactions that are especially driven by the local air–fuel ratio [28–30]. By using advanced SOIs, the HRF ignition delay increases due to the local air–fuel ratio "over leaning", and it generates engine instability as evident from high values of the coefficient of variation (COV) of indicated mean effective pressure (IMEP).

To better highlight differences in combustion RoHR shapes, efficiency, and pollutant emissions in dual-fuel combustion, two SOI sweeps were performed, operating the SCRE at low loads, where HC emissions are exacerbated. To highlight the link between cyclic HC emissions and combustion instability, highly unstable engine operating conditions were specifically studied. Moreover, to better understand the reason behind the NOx production with dual-fuel combustion, different boost pressures and percent energy substitution (PES) of natural gas were considered. The engine operating conditions presented in this paper are summarized in Tables 2 and 3. The experimental studies were performed at a speed of 1339 RPM, which was identified as the "B-speed" of the original PACCAR MX-11 engine and 1000 cycles of data were collected for each operating point. For the analyzed SOI sweeps, the injection pressure was kept at 500 bar, and the PES was maintained at 80% and 75%, respectively. The PES was computed using Equation (1).

$$\text{PES} = \frac{\dot{m}_{\text{LRF}} LHV_{\text{LRF}}}{\dot{m}_{HRF} LHV_{HRF} + \dot{m}_{HRF} LHV_{\text{HRF}}} \quad (1)$$

Table 2. Engine operating conditions that were chosen to study the impact of cyclic variations on HC emissions at Intake Pressure 1.5 bar, Intake Temperature 20 °C, IMEP 5 bar and PES 80%.

SOI [deg aTDC]	Intake Pressure [bar]	Intake Temperature [°C]	PES [%]	COV IMEPg [%]
−10	1.5	20	80	6
−20	1.5	20	80	5.7
−30	1.5	20	80	3.8
−40	1.5	20	80	6.3
−50	1.5	20	80	17.1
−60	1.5	20	80	22

Table 3. Engine operating conditions that were chosen to study the impact of cyclic variations on NOx emissions at Intake Pressure 1.48 bar, Intake Temperature 20 °C, IMEP 5 bar and PES 75%.

SOI [deg aTDC]	Intake Pressure [bar]	Intake Temperature [°C]	PES [%]	COV IMEPg [%]
−10	1.48	20	75	7.4
−15	1.48	20	75	8.2
−20	1.48	20	75	6.3
−25	1.48	20	75	4.8
−30	1.48	20	75	3.6
−35	1.48	20	75	4.7
−40	1.48	20	75	5.9
−45	1.48	20	75	7.6
−50	1.48	20	75	11

To study in depth the combustion process at different engine conditions, the RoHR, the cumulative heat release, and the main combustion indexes, such as IMEP, location (CAD) of 50% heat release (CA50), Peak Pressure Rise Rate (PPRR), were calculated starting from Equation (2) (γ represents the specific heat ratio, V and p are the combustion chamber volume and pressure, respectively, and dV and dp are their derivatives) [31] for each engine cycle:

$$ROHR = \frac{1}{\gamma - 1} \cdot V \cdot \frac{dp}{d\theta} + \frac{\gamma}{\gamma - 1} \cdot p \cdot \frac{dV}{d\theta} \quad (2)$$

The Start of Injection (SOI) was swept from a retarded injection timing of −10 deg aTDC to an advanced injection timing of −50 deg aTDC at constant intake and rail pressures.

Figure 2 shows the impact of the SOI on the average CA50 at a constant IMEP of approximately 5 bar running the sweep described in Table 3. The trend reported in Figure 2 clearly shows the typical behavior during the transition between a typical two-stage profile to a Gaussian profile for dual-fuel combustion in terms of the "switchback" of CA50 direction with respect to the SOI [26,32]. The reason of the SOI-CA50 trend shown in Figure 2 is related to the fact that advancing the angular position of the diesel injection within the cycle increases the ignition delay because of different charge thermodynamic conditions [28,30].

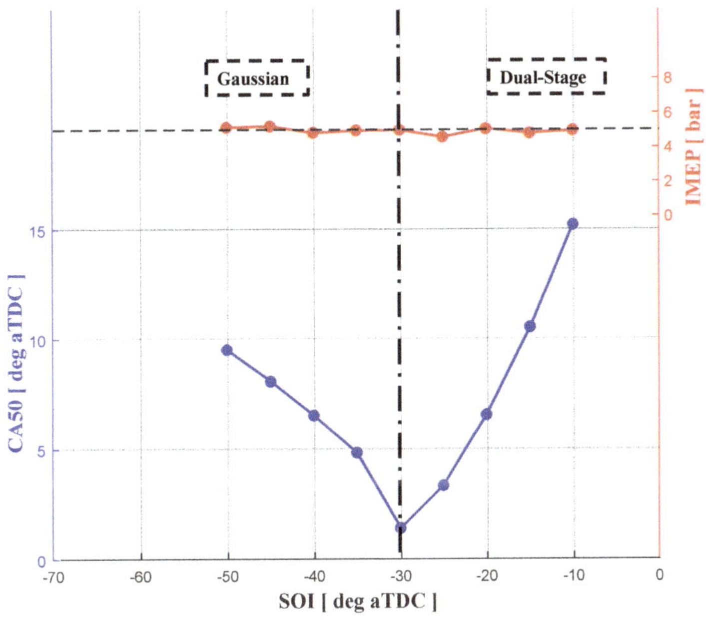

Figure 2. Average CA50 and IMEP measured during the SOI sweep at a constant IMEP of 5 bar.

To better explain the SOI-CA50 trend, the difference in combustion RoHR shapes during the SOI sweeps at the same load (5 bar IMEP) are shown in Figure 3. As evident from this figure, starting from SOI −10 to SOI −30 deg aTDC the combustion shows the typical dual-stage RoHR shape: a first peak in RoHR generated by the combustion of a relatively non-homogeneous air–fuel mixture related to the HRF injection then followed by combustion in the more homogeneous natural gas–air mixture. As has been widely documented in the literature [3], such behavior is generated by very short ignition delay of the HRF when injected close to TDC (the air–fuel mixture pressure, temperature, and the charge stratification are high). After the first combustion phase, the RoHR reported in Figure 3a shows the typical second RoHR stage is characterized by smoother and slower energy release process, due to the combustion of the LRF–air mixture far from the stratified zone. Based on the previous explanation, running the engine with late SOIs, from −30 to −10 deg aTDC, CA50 and SOI follow the same movement direction because of the very short HRF ignition delay and favorable charge thermodynamic conditions (Figure 2, right side).

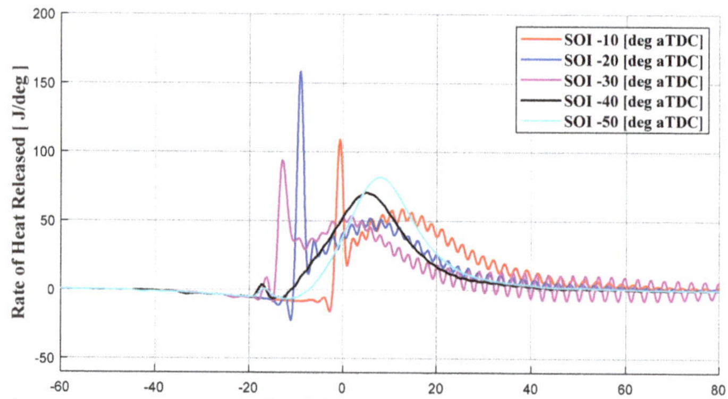

(a)

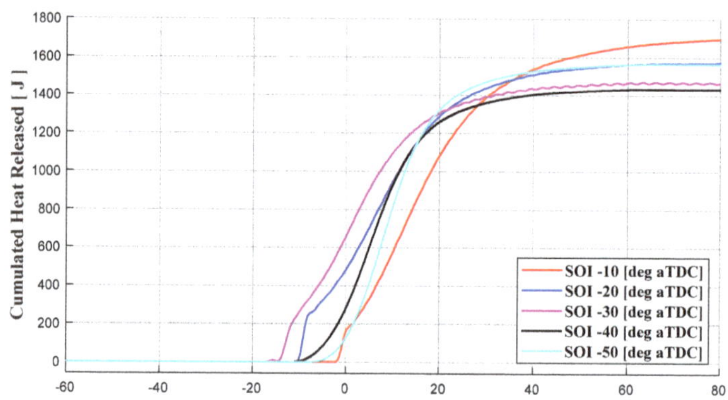

(b)

Figure 3. (**a**) Rate of Heat Release, and (**b**) Cumulate Heat Released curves testing different SOI at the same load (5 bar IMEP), PES (80%), and boost pressure 1.5 bar.

A further advance in SOI, from −30 to −50 deg aTDC, produces a completely different behavior, characterized by a smoother single-stage combustion RoHR that appears "Gaussian" in shape. On the contrary, running the engine with early SOIs, the CA50 retards when the SOI advances (Figure 2, left side). Due to the extremely unfavorable LRF-air mixture thermodynamic conditions when the HRF is injected at early SOIs, the HRF ignition delay increases, allowing for greater HRF mixing with the LRF–air mixture and avoiding the creation of stratified zones close to the injector tip. This enhanced mixing leads to a relatively slower and smoother combustion process characterized by the single-stage Gaussian RoHR curve. It is important to highlight the presence of the Low–Temperature Heat Release (LTHR), approximately at −16 deg aTDC, which represents a well-known phenomenon [33] often visible in single-stage dual-fuel combustion.

Moreover, from Figure 3a, it is important to observe the presence of remarkable RoHR oscillations during the expansion stroke when the engine was run at a retarded SOI, from −10 to −30 deg aTDC. Such a phenomenon is typically triggered by the pressure waves generated by an impulsive combustion process, such as the first stage combustion in Figure 3a [34]. Such a very short and intense energy release increases the amplitude of the resonance frequencies related to the vibrational mode of the combustion chamber. To better analyze this aspect, further analysis is currently being performed.

As has been well documented by the literature [35], the transition from two-stage to Gaussian single-stage combustion produces remarkable differences both in terms of pollutants and efficiency. Figure 3b, which shows the ensemble-average Cumulated Heat Released (CHR) at different SOIs, confirms that even if the CA50 has been varied, the load has been kept almost constant (maximum CHR close to 1500 J). Figure 4 shows the average Indicated Specific Fuel Consumption (ISFC) in terms of diesel-equivalent ISFC and the main pollutant emissions (total HC, CO and NOx) sampled by the Richmond gas analyzer during each test.

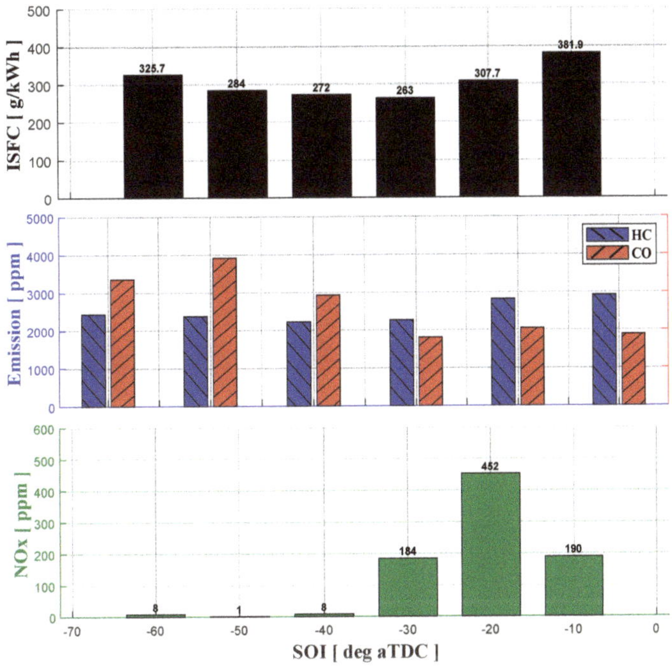

Figure 4. ISFC and pollutants (HC, CO and NOx) production during the SOI sweep running the SCRE engine in dual-fuel mode at low load (IMEP 5 bar), 75% PES, and boost pressure 1.48 bar. Note: For ease of comparison and because the nominal load was held constant at IMEP = 5 bar, steady-state emissions are presented in raw units (ppm) instead of the customary brake-specific units.

From Figure 4, typical ISFC and emissions trends are observed. The increase in NOx is linked to the more locally stratified combustion leading to higher local temperatures, while ISFC and HC reductions are linked to higher combustion efficiencies. It is important to highlight that even if the CA50 was around maximum brake torque (MBT) or maximum IFCE location for standard ICEs (i.e., 5–7 deg aTDC at SOI equal to −30 deg aTDC), the two-stage RoHR and the associated stratified combustion (from SOI −10 to −30 deg aTDC) are characterized by high NOx emissions. As a result, despite its benefit in combustion controllability through the injection position, the two-stage RoHR shows limitations in reaching both goals simultaneously.

Since the aim of dual-fuel combustion is to drop both emissions and fuel consumption as much as possible, further advance in SOI generates a Gaussian combustion and it shows the best tradeoff in pollutants and efficiency. The experimental evidence reported in Figure 4 is well-supported by data available in the dual-fuel literature [26,27]. A smooth combustion characterized by CA50 close to MBT produces very low NOx. A further increment in SOI generates a slight rise in ISFC and HC mainly because the CA50 occurs later as the piston is further down the expansion stroke. Furthermore, the CO generation increases mainly because the CA50 is retarded, the combustion duration increases with a direct impact in lowering the exhaust temperature. Figure 5 shows ISFC and pollutants emissions for engine operation at a slightly higher PES of 80% and boost pressure 1.48 bar. As expected, even at slightly higher PES, SOI variations exhibit similar trends.

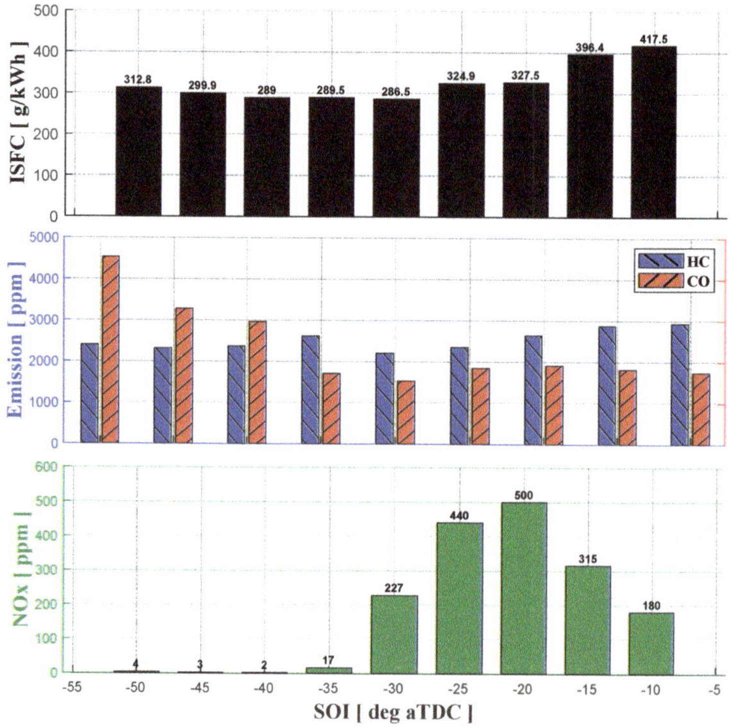

Figure 5. ISFC and pollutants (HC, CO and NOx) production during the SOI sweep running the SCRE engine in dual-fuel mode at low load (IMEP 5 bar), 80% PES, and boost pressure 1.5 bar.

Despite previous efforts in which the authors demonstrated significant reductions in fuel consumption and pollutants by performing a "calibration style" parametric study [17], this work is focused on demonstrating the links between high HC and NOx emissions and the combustion process. Therefore, the focus is specifically not on determining the "best operating point" with dual-fuel combustion.

The following section demonstrates the links between the combustion indices and the information obtained from the high frequency (cycle-resolved) pollutants measurements with low-load dual-fuel combustion.

4. Results and Discussion

To analyze the impact of the combustion characteristics on cycle-resolved pollutants production, attention was focused on those operating conditions that led to high cycle-to-cycle variability (and associated high HC emissions) and to high NOx emissions.

4.1. Cyclic HC—IMEP Correlation

As described in the Introduction, one of the aims of this work is to demonstrate the correlation between the cyclic HC production and IMEP (gross) by using a crank-resolved measurement system operating the heavy-duty SCRE in dual-fuel mode.

Our previous efforts [17] demonstrated that despite remarkable improvements in pollutant emissions and fuel consumption reduction possible with dual-fuel combustion at very advanced SOI, the combustion stability is strongly influenced by the PES and the thermodynamic conditions of the air–fuel mixture. As a result, through the analysis of the experimental data acquired during the performed SOI sweeps, to find out the HC–IMEP correlation it was decided to focus the attention on the 80% PES tests described in Figure 4.

The two engine operating points identified to highlight this correlation are characterized by approximately the same IMEP and CA50 but very different COV of IMEP (we discuss this aspect later in the paper). As reported in Figure 6, the spread of the cyclic IMEP and CA50 are seen by analyzing a subsystem of 300 consecutive cycles (same considerations can be extended to the entire dataset) in the two identified engine conditions at low load: "Advanced SOI" (SOI −60 deg aTDC) and "Retarded SOI" (−10 deg aTDC).

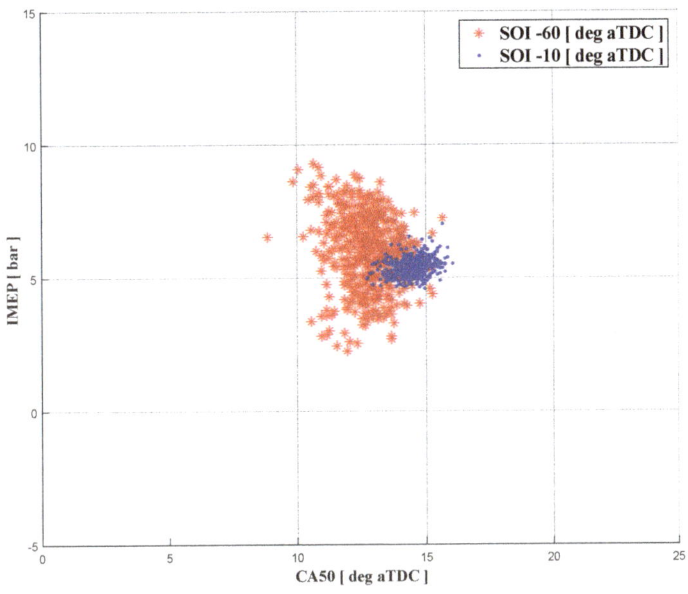

Figure 6. Comparison between measured cyclic CA50-IMEP gross (300 engine cycles) related to two different engine conditions: SOI −60 deg aTDC called "Advanced SOI", and SOI −10 deg aTDC called "Retarded SOI".

For dual-fuel combustion at very advanced SOI, the ignition process (as discussed previously) is driven more by in-cylinder chemistry, which in turn, is affected by the local air–fuel ratio [26–30,36]. To better clarify this aspect, the cyclic IMEP of two groups of 300 consecutive cycles with different SOIs ("Retarded SOI" and "Advanced SOI") are shown in Figure 7. By using advanced SOI, the HRF ignition delay increases due to the "over leaning" of local air–fuel ratios, and this leads to high COV. The spread of CA50 instead is related to the presence of recovery cycles after "near-misfire" engine cycles. After a very low-efficiency combustion, the residuals are mostly composed by unburned air–fuel mixture (fully premixed, after the entire exhaust and the intake stroke). Such extra chemical power given by the residuals increases the engine load anticipating the CA50 of the following cycle ("recovery cycles") consequently. Such behavior is summarized in Figure 8, where the two random groups of engine cycles (the group are highlighted through the yellow box) highlighted in Figure 7 are reported. As explained in the previous section of the paper, due to the different thermodynamic conditions when the HRF is injected, the two extreme SOIs are characterized by extremely different cyclic variations and combustion shape. It is important to mention that, using the AC dynamometer with its associated speed controller (Figure 1), it was possible to investigate extremely unstable engine conditions while keeping the engine speed at a constant value (engine speed fluctuations were mitigated by the dynamometer speed controller).

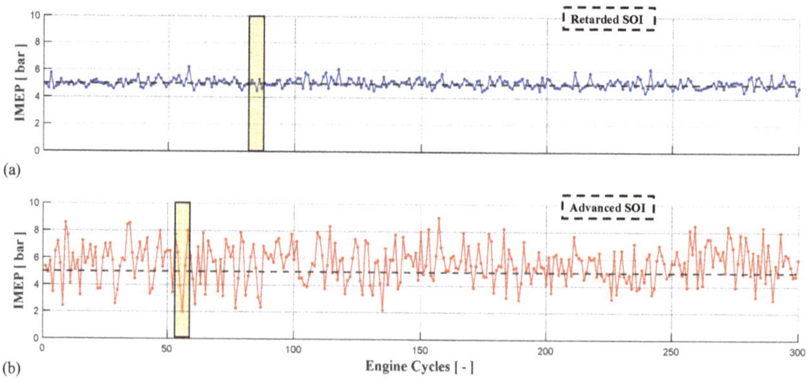

Figure 7. Measured IMEP for 300 consecutive engine cycles: (**a**) "Retarded SOI" condition, (**b**) "Advanced SOI" condition.

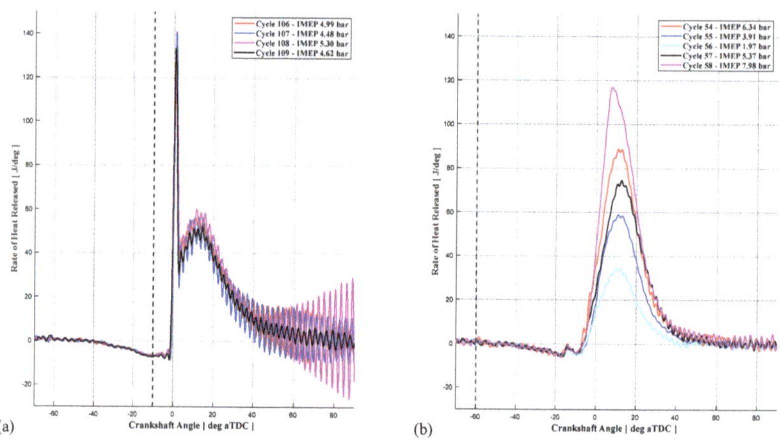

Figure 8. Measured Rate of Heat Released (RoHR) for consecutive cycles in the two different analyzed engine conditions: (**a**) "Retarded SOI", dual-stage combustion, (**b**) "Advanced SOI", "Gaussian" combustion.

As is evident from Figures 7 and 8, due to the different ignition dynamics and consequently combustion RoHR shape and duration, the two analyzed conditions clearly differ in terms of cyclic IMEP behavior: the "Retarded SOI" condition is characterized by very low COV (Figures 7a and 8a) whereas significant IMEP variations are observed for the "Advanced SOI" condition (Figures 7b and 8b).

To demonstrate the impact of the cyclic variations on HC production, cycle-resolved HC emissions were measured using a CAMBUSTION HFR400 fast FID analyzer. By using the fast FID measurement system, which can sample the hydrocarbon emissions at 500 Hz, it was possible to obtain the cycle-resolved HC emissions for each engine cycle. By running the engine at 1339 rpm, the 500 Hz measurement frequency of the fast FID system led to a crank angle resolution of 16 degrees for the cycle-resolved HC measurements. As a result, since the output of the fast FID measurement system was phased with the data acquisition system, it was possible to link the cylinder pressure trace during the combustion process and the respective cycle's HC emissions.

Figure 9 shows the in-cylinder pressure signal for a single engine cycle and the associated fast FID signal. To improve the quality of the fast HC and fast NOx visualization and analysis, a Butterworth low-pass digital filter (300 Hz) has been applied to the FID output.

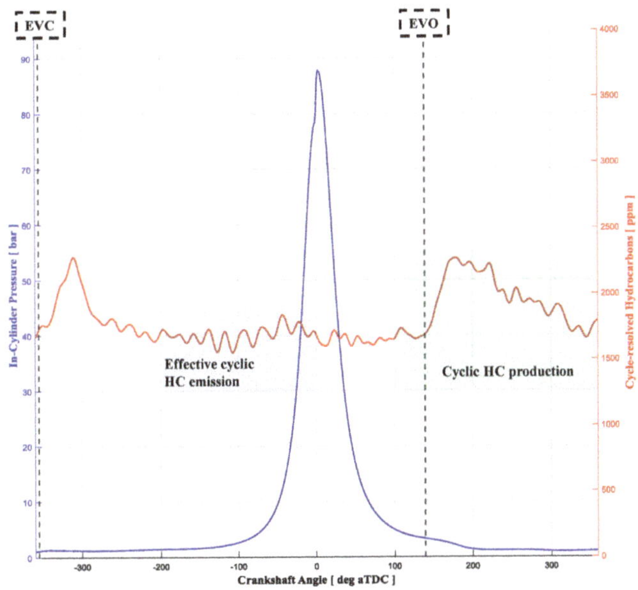

Figure 9. In-cylinder pressure signal and measured cycle-resolved HC emissions for dual-fuel operation at SOI −10 deg aTDC.

It is important to mention that due to the distance between the FID probe and the exhaust port, the fast FID signal is intrinsically characterized by a time-delay (approximately 0.0125 sec in the analyzed conditions) with respect to the exhaust valve opening. Since the analyzed engine operating conditions were run at a constant RPM and under the hypothesis that the exhaust gases pass through the exhaust manifold at the sound speed (i.e., under choked conditions) during the exhaust blowdown process, which can approximately be considered proportional to the square root of the exhaust gas temperature, the time-delay was compensated, and the fast FID signal was synchronized with the exhaust valve opening (EVO) for each cycle. As it can be seen in Figure 9, the instantaneous HC signal starts rising after EVO and, therefore, the information contained in the fast FID signal can be directly correlated with the analyzed cylinder pressure signal.

By the analysis of the instantaneous HC signal (Figure 9), three regions are noticeable, which provide different information about the HC emissions arising from the engine cycle. The first HC peak, located between −360 and −260 deg aTDC, is related to short-circuiting of natural gas from the intake directly to the exhaust. Despite the overlap window of the engine being very small (15 deg), because of the pressure difference between intake and exhaust, a small quantity of LRF–air mixture flows in the cylinder directly to the exhaust, causing the reported peak. The second region can be considered representative of the actual cyclic HC production (green area highlighted in Figure 9) during the combustion process. This area is characterized by the maximum value of instantaneous HC, which is directly related to the HC production during and after the combustion, and a clear descending trend. Such a trend can be reasonably explained through the exhaust gases mixing process in the exhaust manifold during the expansion (the HC concentration lowers progressively). As a result, the cyclic HC production can be evaluated as the average of the fast FID signal during the exhaust stroke (green area in Figure 9). This value represents the "effective cyclic HC emissions" for the following engine cycle. Since the aim of this work is to correlate the IMEP cyclic variations with the HC emissions, the effective cyclic HC emissions was evaluated by averaging the FID signal (blue area highlighted in Figure 9) during the compression stroke, when both intake and exhaust valves are closed. Then, the Net Cyclic HC emission (NCHC), defined by Equation (3), was evaluated as the difference between the cyclic HC production (average of the green area in Figure 9) and the effective cyclic emissions for each engine cycle (average of the blue area in Figure 9).

$$Net\ Cyclic\ HC\ emission\ [ppm] = HC_{cyclic\ production} - HC_{effective\ cyclic\ emission} \quad (3)$$

The *NCHC* represents the net *HC* production (if positive), or the net *HC* reduction (if negative), for the analyzed engine cycle compared to the previous one. As a result, through the NCHC evaluation it is possible to obtain an index that can be easily correlated with the cyclic *IMEP* variation. To evaluate the cyclic *IMEP* variations, the Δ*IMEP*, defined in Equation (4), was calculated as the difference between the *IMEP* of two consecutive engine cycles.

$$\Delta IMEP\ [bar] = IMEP_{n_{th}cycle} - IMEP_{(n-1)_{th}\ cycle} \quad (4)$$

Once *NCHC* and Δ*IMEP* were defined, the cyclic *HC*–Δ*IMEP* correlation was obtained by the comparison between the two engine operating conditions considered in this work: "Retarded SOI" and "Advanced SOI". Figure 10 clearly shows that a net *HC* reduction is related to a more efficient combustion process (positive Δ*IMEP)*. On the other hand, if the combustion decreases in efficiency (negative Δ*IMEP*) the *NCHC* increases. It is interesting to notice that even if the spread of the *IMEP* is extremely different (because of differences in the nature of combustion at the two SOIs as evident from the different RoHR shapes), the correlation between the *NCHC* and the Δ*IMEP* is still valid for both analyzed engine conditions. Therefore, it appears that the obtained linear correlation (R^2 = 0.86) gives a reasonable indication about one the main source of *HC* production in RCCI engines and the relation between the cyclic variability and *HC* emissions in diesel–NG dual-fuel combustion.

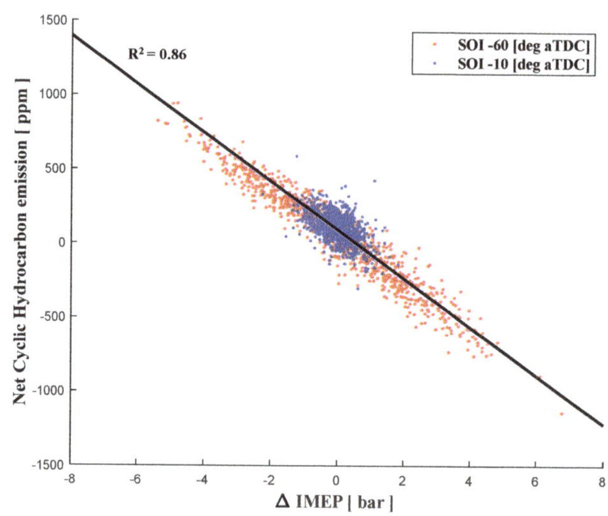

Figure 10. Cyclic HC-IMEP correlation for 300 consecutive cycles in the two different analyzed engine conditions: SOI −60 deg aTDC called "Advanced SOI" and SOI −10 deg aTDC called "Retarded SOI".

4.2. Cyclic NOx–RoHRmax Correlation

As has been widely reported in the literature [27,35], one of the main challenges with adapting dual-fuel combustion for the entire speed–load range production engines is related to the control of the combustion. Despite the fact that Gaussian RoHR shape at advanced SOIs is accompanied by very low NOx and high efficiency, its high sensitivity to in-cylinder thermodynamic conditions and the SOI position significantly reduces the operating range of the engine. As a result, to increase the combustion controllability avoiding knock or misfire, previous efforts proposed retarded SOI operation for a certain number of engine cycles [17,26,35]. By using such SOI actuation, the engine can instantaneously recover from anomalous combustion events, preventing failures or damage to other components, such as the turbocharger, exhaust line, auxiliaries, etc.

Even though retarding the SOI can be considered an effective way to mitigate the well-known dual-fuel combustion, especially knock, moving from a single-stage Gaussian RoHR to a two-stage RoHR increases the NOx emission. As reported in Figure 3a and confirmed by pollutants emissions shown in Figure 4, the two-stage combustion (SOI from −10 to −30 deg aTDC) is characterized by a very sharp first combustion stage and high NOx production. As mentioned before, such behavior is generated by the very short ignition delay of the HRF when injected close to TDC (pressure, temperature and the stratification of the air–fuel mixture are high) which ignites the amount of charge in the stratified zone.

Since NOx formation is driven by local temperatures potentially caused by the local air–fuel ratio stratification in dual-fuel combustion, RoHR calculated using a global combustion model is not directly referable to NOx emissions. This is made even more difficult especially when the nature of combustion changes significantly (i.e., from two-stage to single-stage Gaussian RoHR).

However, because the NOx production is promoted when a significant amount of energy is released in a short time, the link between the combustion behavior and the measured NOx can be obtained by examining combustion indices related to knock. Typically, such information can be easily obtained by the analysis of the RoHR, and in particular its maximum value.

From Figure 3a, it is evident that for combustion characterized by two-stage RoHR (SOI from −10 to −30 deg aTDC), the maximum value of the RoHR (RoHRmax), typically occurring in the first stage, is significantly and consistently higher with respect to the more advanced SOIs (SOI from −40 to −50 deg aTDC). Figure 11 shows the steady-state (slow-speed) NOx emissions measured with the Richmond bench and the RoHRmax (300 cycles) during the SOI sweep. It is evident that both NOx and RoHRmax follow the same trend, demonstrating a potential link between the identified combustion parameter and NOx production.

Focusing the attention on the SOI range characterized by two-stage combustion (SOI from −10 to −30 deg aTDC), because the first combustion stage progressively becomes more intense (Figure 2a), both NOx and RoHRmax increase. The reported trend is generated by the decreasing HRF ignition delay, which promotes local air–fuel ratio stratification and higher local temperatures inside the combustion chamber. A further advancement in SOI (from −35 to −50 deg aTDC) modifies the RoHR shape, characterized by a smoother energy release (Gaussian profile). Such behavior is mainly related to the HRF ignition delay increment because of the unfavorable thermodynamic conditions in the combustion chamber when the HRF is injected [28,29]. The very low NOx emissions shown by Figure 11 at advanced SOIs confirms a potential link between the combustion shape (i.e., the existence of a distinct first stage of RoHR) and NOx emissions. It is important to highlight that the RoHRmax tends to increase when SOI is advanced because the center of combustion moves to MBT (Figure 2). A further confirmation of this trend can be found by the information shown in Figure 4, where ISFC and pollutants are reported.

Upon identification of the link between RoHRmax and NOx emissions using the ensemble-averaged RoHR data and slow-speed NOx measurements, cycle-to-cycle analysis was performed.

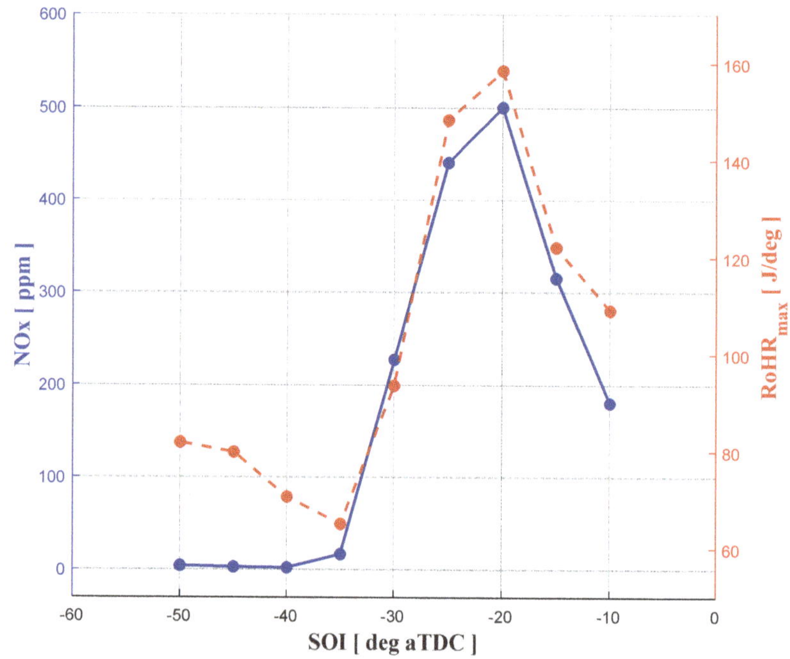

Figure 11. Average NOx-RoHRmax correlation obtained through slow speed pollutant measurement (Richmond measurement system) running a SOI sweep in dual-fuel mode at low-load.

Following the same approach described in the previous section for identifying the HC–IMEP correlation, the fast NOx signal was acquired and synchronized with the EVO (after compensating for time-delay as discussed above). By phasing the instantaneous NOx emissions with the exhaust stroke, it was possible to clearly identify the NOx emissions associated with each engine cycle. Since the signal output characteristics of the CAMBUSTION CLD500 have the same features as the fast HC analyzer, the same signal processing (acquisition and filtering) was adopted. Figure 12 shows the CAMBUSTION CLD500 signal (red line) and in-cylinder pressure signal (blue line) of an engine cycle characterized by high NOx emission (retarded SOI −20 deg aTDC).

Figure 12 clearly shows the peak of NOx emissions placed at the end of expansion stroke (EVC −355 deg aTDC) which can be assumed as the cyclic NOx production. Such behavior can be related to the delay of the Fast NOx before starting to measure the exhaust gases of the actual cycle. As documented by Peckham et al. [37] and Schurov et al. [38], the evident instantaneous NOx signal drop reported in Figure 12, around EVO, is related to the residence time of the exhaust gases behind the valve. The first portion of the gas sampled will be that which was released at the end of the previous exhaust stroke and then stayed in the port throughout the entire exhaust valve closed period resulting in a low NOx concentration. Moreover, since fast NOx and fast FID data have the same characteristics in term of signal output, the cycle resolved data have a crank angle resolution of 16 degrees that contributes retarding the rise of NOx signal.

With the aim of confirming the NOx-RoHRmax correlation identified through the analysis of the average data in Figure 11, the two stage RoHR combustion region (SOI from −10 to −30 deg aTDC) was considered. As clearly shown in Figure 13, where a cycle-to-cycle comparison (300 consecutive engine cycles) between the maximum value of the instantaneous NOx and the calculated RoHRmax was performed, the same trend reported in Figure 11 can be noticed. As discussed before, by advancing the SOI, the magnitude of the first RoHR stage initially increases and subsequently decreases, with obvious consequences in terms of NOx production. A further advancement in SOI generates near-zero NOx production because of the different combustion behavior (Gaussian RoHR shape), confirming the link between the combustion RoHR shape and NOx production. This strong linear correlation ($R^2 = 0.82$) between NOx emissions and the transformation of the RoHR shape in dual-fuel combustion has been previously observed by several researchers, most recently by Partridge et al. [32], who attributed the RoHR shape transformation to local air–fuel ratio stratification.

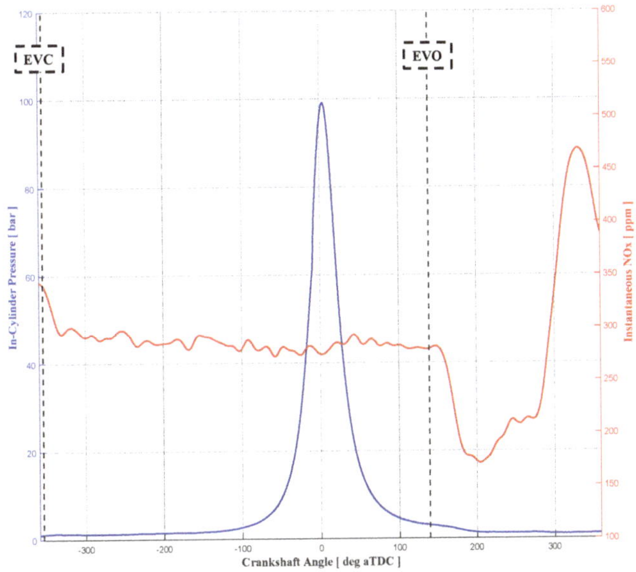

Figure 12. In-cylinder pressure signal and Instantaneous NOx production for a dual-fuel engine operating condition at SOI −20 deg aTDC.

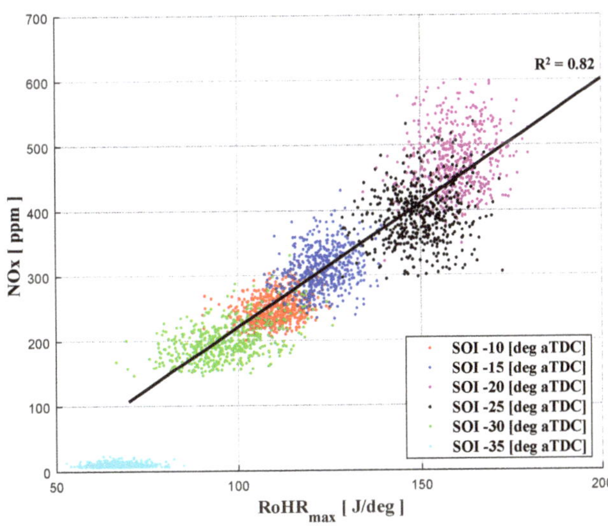

Figure 13. Cycle-to-cycle NOx-RoHRmax correlation obtained through high-frequency pollutant measurement (CAMBUSTION measurement system) running a SOI sweep in dual-fuel mode at low load.

5. Conclusions

This work highlights the correlation between combustion characteristics and pollutants emissions, especially unburned hydrocarbons (HC) and oxides of nitrogen (NOx) based on fast FID and fast NOx measurements acquired cycle-by-cycle. The experimental investigation was performed on a Single-Cylinder Research engine run in dual-fuel diesel–natural gas mode at a fixed speed and a fixed low load of 5 bar IMEP gross and 1339 rpm, respectively. The natural gas PES was kept constant at 80% and the diesel was injected at an injection pressure of 500 bar, while the HRF SOI was varied from −10 to −60 deg aTDC.

The instantaneous pollutants emissions were acquired using CAMBUSTION HFR400 and CLD 500 with sampling frequency of 500 Hz able to obtain cycle-resolved data. To further validate the values coming from the CAMBUSTION instrumentations, a standard five-gas emission bench was used to compare the outputs in terms of average (steady-state) pollutants concentrations from different sources. Once compensated for the time-delay of the fast emission measurement systems with respect to the actual engine cycle (due to the distance between the exhaust valve and the sampling probe), the analysis of the acquired data allowed to define two correlations aimed at demonstrating the sources of each pollutants produced.

The correlation between cycle-to-cycle variability and the HC production were identified by comparing two different engine operating conditions of −10 deg aTDC and −60 deg aTDC, called "Retarded SOI" and "Advanced SOI", respectively. Introducing the Net Cyclic Hydrocarbon (NCHC) emissions as representative of the net HC production (if positive), or net HC reduction (if negative), for the analyzed engine cycle compared to the previous one, the cyclic variability was defined through the calculation of the IMEP difference between of two consecutive engine cycles (ΔIMEP). Finally, a linear experimental correlation ($R^2 = 0.86$) between NCHC and ΔIMEP was obtained, providing a reasonable physical explanation for high HC emission running diesel–NG dual-fuel combustion at low load with advanced injection timing.

Following the same approach, a physical explanation for high NOx emissions in diesel–NG dual-fuel combustion was proposed. Through the analysis of the ensemble-averaged RoHR data and slow-speed measurements, a potential link between the combustion shape (i.e., the existence of a distinct first stage of RoHR, summarized by the RoHRmax) and NOx emissions was identified. The analysis of the cycle-to-cycle combustion data and crank-resolved NOx measurements confirmed the existence of a clear trend through the identification of a linear correlation ($R^2 = 0.82$) between the maximum value of the instantaneous NOx and the calculated RoHRmax. Since the analysis performed in this paper consider only few SOIs at a constant load (low load), further activities are currently being performed to verify the obtained correlation at different engine operating conditions of RPM, load, intake pressure, rail pressure and number of HRF injections.

Author Contributions: Conceptualization, G.S., K.K.S. and S.R.K.; methodology, G.S. and A.N. and F.P.; investigation, G.S., A.N., N.C. and S.R.K.; resources, G.S. and A.N.; data curation, G.S.; writing—original draft preparation, G.S. and V.R. writing—review and editing, K.K.S., S.R.K., F.P. and P.P.; visualization, G.S. and V.R.; supervision, F.P., S.R.K. and K.K.S.; project administration, V.R., F.P., S.R.K. and K.K.S. All authors have read and agreed to the published version of the manuscript.

Funding: This paper was developed based on partial funding from the Alliance for Sustainable Energy, LLC, Managing and Operating Contractor for the National Renewable Energy Laboratory for the U.S. Department of Energy (Prime Award #DE-AC36-08GO28308; Subaward #NHQ-9-82305-01). Additional financial support was provided by the Department of Mechanical Engineering of The University of Alabama, and the University of Bologna.

Data Availability Statement: The data presented in this paper will be provided by the authors on request.

Acknowledgments: The authors acknowledge funding from the Alliance for Sustainable Energy, LLC, Managing and Operating Contractor for the National Renewable Energy Laboratory for the U.S. Department of Energy as well as from the University of Bologna and The University of Alabama.

Conflicts of Interest: The authors declare no conflict of interest.

Abbreviations

Ac	Alternate current
aTDC	after Top Dead Center
bTDC	before Top Dead Center
CAD	Crank Angle Degree
CDC	Conventional Diesel Combustion
CO	Carbon Oxides
COV imep	Coefficient of Variation of IMEP
dV	Derivative Volume
dp	Derivative Pressure
EVC	Exhaust Valve Closing angle
EVO	Exhaust Valve Opening angle
FCE	Fuel Conversion Efficiencies
HCS	Unburned Hydrocarbons
HCCI	Homogeneous Charge Compression Ignition
HP	Horse Power
HRF	High-Reactivity Fuel
IMEP	Indicated Mean Effective Pressure
ISFC	Indicated Specific Fuel Consumption
IVC	Intake Valve Closing angle
IVO	Intake Valve Opening angle
LHV_{LRF}	Low Heating Value of Low Reactivity Fuel
LHV_{HRF}	Low Heating Value of High Reactivity Fuel
LRF	Low-Reactivity Fuel
LTC	Low-Temperature Combustion
LTHR	Low-Temperature Heat Release
MBT	Maximum Brake Torque
NCHC	Net Cyclic Hydrocarbons
NG	Natural Gas
NOx	Oxides of Nitrogen
P	Pressure
PES	Percentage Energy Substitution
PREMIER	Premixed Mixture Ignition in the End-gas Region
POMDME	Polyoxymethylene di-methyl ether
RCCI	Reactivity Controller Compression Ignition
RoHR	Rate of Heat Release
RoHRmax	Maximum Rate of Heat Release
SOI	Start of Injection angle
TDC	Top Dead Center
V	Volume
$\Delta IMEP$	Cyclic IMEP variation
$\dot{m}_{HRF}$	High-Reactivity Fuel Flowrate
$\dot{m}_{LRF}$	Low-Reactivity Fuel Flowrate
γ	Adiabatic index

References

1. Thring, R. *Homogeneous-Charge Compression-Ignition (HCCI) Engines*; SAE Technical Paper 892068; SAE: Warrendale, PA, USA, 1989. [CrossRef]
2. Christensen, M.; Johansson, B. *Supercharged Homogeneous Charge Compression Ignition (HCCI) with Exhaust Gas Recirculation and Pilot Fuel*; SAE Technical Paper 2000-01-1835; SAE: Warrendale, PA, USA, 2000. [CrossRef]
3. Imamoto, T.; Kawahara, N.; Tomita, E. PREMIER combustion characteristics of a pilot fuel-ignited dual-fuel biogas engine with consideration of cycle-to-cycle variations. *Fuel* **2022**, *314*, 123049. [CrossRef]

4. Azimov, U.; Tomita, E.; Kawahara, N.; Harada, Y. Premixed mixture ignition in the end-gas region (PREMIER) combustion in a natural gas dual-fuel engine: Operating range and exhaust emissions. *Int. J. Engine Res.* **2011**, *12*, 484–497.
5. Shin, J.; Choi, J.; Seo, J.; Park, S. Pre-chamber combustion system for heavy-duty engines for operating dual fuel and diesel modes. *Energy Convers. Manag.* **2022**, *255*, 115365. [CrossRef]
6. Salahi, M.M.; Esfahanian, V.; Gharehghani, A.; Mirsalim, M. Investigating the reactivity controlled compression ignition (RCCI) combustion strategy in a natural gas/diesel fueled engine with a pre-chamber. *Energy Convers. Manag.* **2017**, *132*, 40–53. [CrossRef]
7. Kokjohn, S.; Hanson, R.; Splitter, D.; Kaddatz, J.; Reitz, R. Fuel Reactivity Controlled Compression Ignition (RCCI) Combustion in Light- and Heavy-Duty Engines. *SAE Int. J. Engines* **2011**, *4*, 360–374.
8. Hanson, R.; Kokjohn, S.; Splitter, D.; Reitz, R. Fuel Effects on Reactivity Controlled Compression Ignition (RCCI) Combustion at Low Load. *SAE Int. J. Engines* **2011**, *4*, 394–411.
9. Paykani, A.; Garcia, A.; Shahbakhti, M.; Rahnama, P.; Reitz, R.D. Reactivity controlled compression ignition engine: Pathways towards commercial viability. *Appl. Energy* **2021**, *282*, 116174.
10. Benajes, J.; García, A.; Monsalve-Serrano, J.; Villalta, D. Exploring the limits of the reactivity controlled compression ignition combustion concept in a light-duty diesel engine and the influence of the direct-injected fuel properties. *Energy Convers. Manag.* **2018**, *157*, 277–287.
11. Kakaee, A.; Rahnama, P.; Paykani, A. Numerical Study of Reactivity Controlled Compression Ignition (RCCI) Combustion in a Heavy-Duty Diesel Engine Using 3D-CFD Coupled with Chemical Kinetics. *Int. J. Automot. Eng.* **2014**, *4*, 792–804.
12. Reitz, R.D.; Duraisamy, G. Review of high efficiency and clean reactivity controlled compression ignition (RCCI) combustion in internal combustion engines. *Prog. Energy Combust. Sci.* **2015**, *46*, 12–71.
13. Krishnan, S.; Srinivasan, K.; Singh, S.; Bell, S.; Midkiff, K.; Gong, W.; Fiveland, S.; Willi, M. Strategies for Reduced NOx Emissions in Pilot-Ignited Natural Gas Engines. *ASME. J. Eng. Gas Turbines Power* **2004**, *126*, 665–671.
14. Raihan, M.; Guerry, E.; Dwivedi, U.; Srinivasan, K.; Krishnan, S. Experimental Analysis of Diesel-Ignited Methane Dual-Fuel Low-Temperature Combustion in a Single-Cylinder Diesel Engine. *J. Energy Eng.* **2014**, *141*, C4014007.
15. Krishnan, S.; Srinivasan, K.; Raihan, M. The effect of injection parameters and boost pressure on diesel-propane dual fuel low temperature combustion in a single-cylinder research engine. *Fuel* **2016**, *184*, 490–502.
16. Narayanan, A.; Hariharan, D.; Partridge, K.; Pearson, A.; Srinivasan, K.; Krishnan, S. Impact of Low Reactivity Fuel Type and Energy Substitution on Dual Fuel Combustion at Different Injection Timings. *Energies* **2023**, *16*, 1807. [CrossRef]
17. Hariharan, D.; Partridge, K.; Narayanan, A.; Srinivasan, K. Strategies for Reduced Engine-Out HC, CO, and NOx Emissions in Diesel-Natural Gas and POMDME-Natural Gas Dual-Fuel Engine. *SAE Int. J. Adv. Curr. Prac. Mobil.* **2022**, *4*, 1264–1278.
18. Selim, M.Y.E. Effect of engine parameters and gaseous fuel type on the cyclic variability of dual fuel engines. *Fuel* **2005**, *84*, 961–971.
19. Ozdor, N.; Dulger, M.; Sher, E. *Cyclic Variability in Spark Ignition Engines A Literature Survey*; SAE Technical Paper 940987; SAE: Warrendale, PA, USA, 1994.
20. Hamai, K.; Kawajiri, H.; Ishizuka, T.; Nakai, M. *21st Symposium (International) on Combustion*; Technical University of Munich West Germany: München, Germany, 1986.
21. Galloni, E. Analyses about parameters that affect cyclic variation in a spark ignition engine. *Appl. Therm. Eng.* **2008**, *29*, 1131–1137. [CrossRef]
22. Matekunas, F. *Modes and Measures of Cyclic Combustion Variability*; SAE Technical Paper 830337; SAE: Warrendale, PA, USA, 1983.
23. Srinivasan, K.K.; Agarwal, A.K.; Krishnan, S.R.; Mulone, V. *Natural Gas Engines—For Transportation and Power*; Springer: Berlin/Heidelberg, Germany, 2019.
24. Jha, P.R.; Krishnan, S.R.; Srinivasan, K.K. Impact of methane energy fraction on emissions, performance and cyclic variability in low-load dual fuel combustion at early injection timings. *Int. J. Engine Res.* **2021**, *22*, 1255–1272.
25. Cheng, Q.; Ahmad, Z.; Kaario, O.; Martti, L. Cycle-to-cycle variations of dual-fuel combustion in an optically accessible engine. *Appl. Energy* **2019**, *254*, 113611.
26. Ravaglioli, V.; Carra, F.; Moro, D.; De Cesare, M.; Stola, F. *Remote Sensing Methodology for the Closed-Loop Control of RCCI Dual Fuel Combustion*; SAE Technical Paper 2018-01-0253; SAE: Warrendale, PA, USA, 2018. [CrossRef]
27. Curran, S.; Hanson, R.; Wagner, R.; Reitz, R. *Efficiency and Emissions Mapping of RCCI in a Light-Duty Diesel Engine*; SAE Technical Paper 2013-01-0289; SAE: Warrendale, PA, USA, 2013. [CrossRef]
28. Matsuura, K.; Iida, N. *Effect of Temperature-Pressure Time History on Auto-Ignition Delay of Air-Fuel Mixture*; SAE Technical Paper 2018-01-1799; SAE International: Warrendale, PA, USA, 2018. [CrossRef]
29. Silvagni, G.; Ravaglioli, V.; Falfari, S.; Ponti, F.; Mariani, V. Development of a Control-Oriented Ignition Delay Model for GCI Combustion. *Energies* **2022**, *15*, 6470. [CrossRef]
30. Ravaglioli, V.; Ponti, F.; Silvagni, G.; De Cesare, M. Development of a Methodology for the Investigation of Residual Gases Effects on Gasoline Compression Ignition. In Proceedings of the ASME 2020 Internal Combustion Engine Division Fall Technical Conference, Virtual, Online, 4–6 November 2020. [CrossRef]
31. Heywood, J. *Internal Combustion Engine Fundamentals*; McGraw-Hill Education: New York, NY, USA, 1988.
32. Partridge, K.; Jha, P.; Srinivasan, K.; Krishnan, S. An experimental and computational analysis of combustion heat release transformation in dual fuel combustion. *Fuel* **2023**, *341*, 127561. [CrossRef]

33. Shibata, G.; Oyama, K.; Urushihara, T.; Nakano, T. *Correlation of Low Temperature Heat Release with Fuel Composition and HCCI Engine Combustion*; SAE Technical Paper 2005-01-0138; SAE: Warrendale, PA, USA, 2005. [CrossRef]
34. Novella, R.; Pla, B.; Bares, P.; Jiménez, I. Acoustic characterization of combustion chambers in reciprocating engines: An application for low knocking cycles recognition. *Int. J. Engine Res.* **2022**, *23*, 120–131. [CrossRef]
35. Ravaglioli, V.; Ponti, F.; Carra, F.; Corti, E.; De Cesare, M.; Stola, F. Combustion Indexes for Innovative Combustion Control. *SAE Int. J. Engines* **2017**, *10*, 2371–2381. [CrossRef]
36. Del Vescovo, D.; Kokjohn, S.; Reitz, R. The Effects of Charge Preparation, Fuel Stratification, and Premixed Fuel Chemistry on Reactivity Controlled Compression Ignition (RCCI) Combustion. *SAE Int. J. Engines* **2017**, *10*, 1491–1505. [CrossRef]
37. Peckham, M.; Hands, T.; Burrell, J.; Collings, N.; Schurov, S. *Real Time In-Cylinder and Exhaust NO Measurements in a Production SI Engine*; SAE Technical Paper 980400; SAE: Warrendale, PA, USA, 1998. [CrossRef]
38. Schurov, S.; Collings, N.; Hands, T.; Peckham, M.; Burrell, J. *Fast Response NO/HC Measurements in the Cylinder and Exhaust Port of a DI Diesel Engine*; SAE Technical Paper 980788; SAE: Warrendale, PA, USA, 1998. [CrossRef]

MDPI
St. Alban-Anlage 66
4052 Basel
Switzerland
www.mdpi.com

Energies Editorial Office
E-mail: energies@mdpi.com
www.mdpi.com/journal/energies

www.ingramcontent.com/pod-product-compliance
Lightning Source LLC
Chambersburg PA
CBHW061558210926
51850CB00009B/2142